Business Statistics with Solutions in R

Business Statistics with Solutions in R

Mustapha Abiodun Akinkunmi

DE

G

PRESS

Mustapha Abiodun Akinkunmi
American University of Nigeria
Yola, Nigeria

ISBN: 978-1-5474-1746-9
e-ISBN (PDF): 978-1-5474-0135-2
e-ISBN (EPUB): 978-1-5474-0137-6

Library of Congress Control Number: 2019949179

Bibliographic information published by the Deutsche Nationalbibliothek
The Deutsche Nationalbibliothek lists this publication in the Deutsche Nationalbibliografie;
detailed bibliographic data are available on the Internet at http://dnb.dnb.de.

To My Son
Omar Olanrewaju Akinkunmi

Contents

1 Introduction to Statistical Analysis

The science of statistics supports making better decisions in business, economics, and other disciplines. In addition, it provides necessary tools to summarize data, analyze data, and make meaningful conclusions to arrive at these better decisions. This book is a bit different from most statistics books in that it will focus on using data to help you do business—the focus is not on statistics for statistics' sake but rather on what you need to know to really *use* statistics.

To help in that endeavor, examples will include the use of the R programming language which was created specifically to give statisticians and technicians a tool to create solutions to standard techniques as well as custom solutions.

Statistics is a word originating from the Italian word *stato* meaning "state," and *statista* is an individual saddled with the tasks of the state. Thus, statistics is the collection of useful information to the statista. Its application began in Italy during the 16th century and dispersed to other countries of the world. At present, statistics covers a wide range of information in every aspect of human activity. It is not limited to the collection of numerical information but includes data summarization, analysis, and presentation in meaningful ways.

Statistical analysis is mainly concerned with how to make generalizations from data. Statistics is a science that deals with information. In order to perform statistical analysis on information (data) you have on hand or that you collect, you may need to transform the data or work with the data to get it into a form where it can be analyzed using statistical techniques. Information can be found in *qualitative* or *quantitative* form. In order to explain the difference between these two types of information, let's consider an example. Suppose an individual intends to start a business based on the information in Table 1.1. Which of the variables are quantitative and which are qualitative? The product price is a quantitative variable because it provides information based on quantity—the product price in dollars. The number of similar businesses and the rent for business premises are also quantitative variables. The location used in establishing the business is a qualitative variable since it provides information about a quality (in this case a location, Nigeria or South Korea). The presence of basic infrastructure requires a (Yes or No) response; these are also qualitative variables.

A quantitative variable represents a number for which arithmetic operations such as averaging make sense. A qualitative (or categorical) variable is concerned with a decision. In a case where a number is applied to separate members of different categories of a qualitative variable, the assigned number is subjective

https://doi.org/10.1515/9781547401475-001

Table 1.1: Business feasibility data.

Product price	Number of similar businesses	Rent for the business premise	Location	Presence of basic infrastructure
US$150	6	US$2000	Nigeria	No
US$100	18	US$3000	South Korea	Yes

but generally intentional. An aspect of statistics is that it is concerned with measurements—some quantitative and others qualitative. Measurements provide the real numerical values of a variable. Qualitative variables can be represented with numbers as well, but such a representation may be arbitrary but useful for the purposes at hand. For instance, you can assign numerics to an instance of a qualitative variable such as Nigeria = 1, and South Korea = 0.

1.1 Scales of Measurement

In order to use statistics effectively, it is helpful to view data in a slightly different way so that it can be analyzed successfully. In performing a statistical test, it is important that all data be converted to the same scale. For example, if your quantitative data is in meters and feet, you need to choose one and convert any data to that scale. However, in statistics there is another definition for *scales of measurement*. Scales of measurement are commonly classified into four categories, namely: nominal scale, ordinal scale, interval scale, and ratio scale.

1. **Nominal Scale:** In this scale, numbers are simply applied to label groups or classes. For instance, if a dataset consists of male and female, we may assign a number to them such as 1 for male and 2 for female. In this situation, the numbers 1 and 2 merely denote the category in which a data point belongs. The nominal scale of measurement is applied to qualitative data such as gender, geographic classification, race classification, and so on.
2. **Ordinal Scale:** This scale allows data elements to be ordered based on their relative size or quality. For example, buyers can rank three products by assigning them 1, 2, and 3, where 3 is the best and 1 is the worst. The ordinal scale does not provide information on how much better one product is compared to others, only that it is better. This scaling is used for many purposes—such as grading, either stanine (1 to 9) or A to F (no E) where a 4.0 is all As and all Fs are 0.0; or rankings, such as on Amazon (1 to 5

stars, so that that 5.0 would be the highest ranking), or on restaurants, hotels, and other data sources which may be ranked on a four- or five-star basis. It is therefore quite important to know the data range when using this data. The availability of this type of data on the internet makes this a good category for experimenting with real data.

3. **Interval Scale:** Interval scales are numerical scales in which intervals have the same interpretation. For instance, in the case of temperature, we have a Celsius scale where water freezes at 0 degrees and boils at 100 degrees and a Fahrenheit scale where water freezes at 32 degrees and 212 degrees is the temperature at which it boils. These arbitrary decisions can make it difficult to take ratios. For example, it is not correct to say 10:00 p.m. is twice as long as 5:00 p.m. However, it is possible to have ratios of intervals. The interval between 2:00 p.m. and 10:00 p.m., which represents a duration of 8 hours, is twice as long as the interval between 2:00 p.m. and 6:00 p.m., which represents a duration of 4 hours. It is important that you understand your data and make sure it is converted appropriately so that in cases like this, interval scale can be used in statistical tests.

4. **Ratio Scale:** This scale allows us to take ratios of two measurements if the two measurements are quantitative in nature. The zero in the ratio scale is an absolute zero. For instance, money is measured in a ratio scale. A sum of US$1000 is twice as large US$500. Other examples of the ratio scale include the measurements of weight, volume, area, or length.

1.2 Data, Data Collection, and Data Presentation

A dataset is defined as a set of measurements gathered on some variables. For example, dividend measurements for 10 companies may be used as a dataset. In this case, the concerned variable is dividend and the scale of measurement is a ratio scale. A dividend for one company may be twice another company's dividend. In the actual observations of the companies' dividends, the dataset might record US$4, US$6, US$5, US$3, US$2, US$7, US$8, US$1, US$10, US$9.

Different approaches are used to collect data. Sometimes a dataset includes the entire population of interest, or it might constitute a sample from the larger population. In the latter situation, the data needs to be carefully collected since we want to draw inferences about the larger population based on the sample. The conclusion made about an entire population by relying on the information in a sample of the population is referred to as a *statistical inference*. This book assigns great importance to the topic of statistical inference (it is fully explained in Chapter 6). The accuracy of statistical inference depends on how data is drawn from the population of interest. It is also critically important to ensure that every segment of the population is sufficiently and equally captured in the sample. Data collection from surveys or experiments has to be appropriately constructed, as statistical inference is based on them.

This subsection focuses on the processing, summarization, and presentation of data and represents the first stage of statistical analysis. What are the reasons behind more attention on inference and population? Is it sufficient to observe data and interpret it? Data inspection is appropriate when the interest is aimed at particular observations, but meaningful conclusions require statistical analysis. Any business decision based on a sample requires statistical analysis to draw meaningful insights. Therefore, the role of statistics in any business entity is both a necessary and sufficient condition for effective and efficient decision-making. For instance, marketing research that aims to examine the link between advertising and sales might obtain a dataset of randomly chosen sales and advertising figures for a given firm. However, much more useful information can be concluded if implications about the process of sample selection and the process involved in that selection are considered. Other businesses like banking might have interest in evaluating the diffusion of a particular model of automated teller machines; a pharmaceutical firm that intends to market its new drug might require proving the drug does not cause negative side effects. These different aims of business entities can be attained through the use of statistical analysis.

Statistics can also have a dark side. The use of statistics takes a significant role in the success of many business activities. Therefore, it is in a company's interest to use statistics to their benefit. So, AIG might say that since 2010 their stock has gone up X%, having been bailed out the year before. You see mutual funds brag all the time about their record since a certain date. You can be certain that right before that date something bad happened. A company may claim that their survey showed that 55% preferred their brand of tomato soup. How many surveys did they conduct? How many different types of soup did they survey where the soup did poorly in statistical results? Another example is that in a test, one aspect of a product can come out well, but other aspects may get bad results, so the company presents the positive results. Sampling is another area where it is easy to cheat and that can compromise the integrity of existing data. There are many other ways to "lie" with statistics and it has become so commonplace that there are books written on the subject. The bottom line is to have a healthy skepticism when you hear statistics promoted in the media.

1.3 Data Groupings

Data is information that captures the qualitative or quantitative features of a variable or a set of variables. Put differently, data can be defined as any set of information that reasonably describes or represents the characteristics of a given entity.

1.3.1 Categories of Data Groupings

In statistics, data can be categorized into *grouped* and *ungrouped* form. Ungrouped data is data in the raw form. For instance, a general list of business firms that you know is raw form data. Grouped data is data that has been organized into groups known as *classes*. This implies that grouped data is no longer in the raw format.

A *class* can be defined as a group of data values within specific group boundaries. A *frequency distribution table* is a table that shows the different measurement categories and how often each category occurs. Plotting such a frequency distribution of grouped data is called as a histogram. A *histogram* is a chart comprised of bars of different heights. The height of each bar depicts the frequency of values in the class represented by the bar. Adjacent bars share sides. It is used only for measured or ordinal data. This is an appropriate way of plotting the frequencies of grouped data. Table 1.2 shows a frequency table of a random sample of 30 small-scale businesses with their number of years of existence.

Table 1.2: Frequency table for grouped data.

Years of Establishment	Number of Businesses
0–2	12
3–5	6
6–8	8
9–11	3
12–14	1

The graph in Figure 1.1 shows the histogram representation for the given data in Table 1.2.

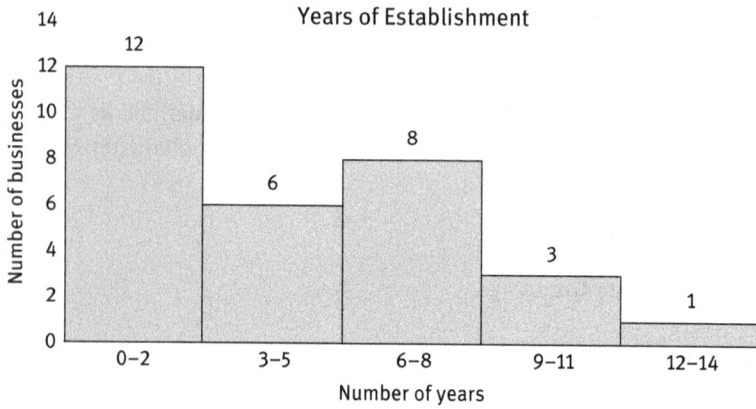

Figure 1.1: Histogram example.

A *data class* refers to a group of data that is similar based on a user-defined property. For instance, ages of students can be grouped into classes such as those in their teens, twenties, thirties, and so on. Each of these groups is known as a *class*. Each class has a specific width known as the class interval or class size. The class interval is crucial in the construction of histograms and frequency diagrams. Class size depends on how the data is grouped. The class interval is usually a whole number. Table 1.3 represents the grouped data that has equal class interval.

Table 1.3: Grouped data with the same class interval.

Daily Income (US$)	Frequency	Class Interval
0–9	5	10
10–19	8	10
20–29	12	10
30–39	17	10
40–49	3	10
50–59	4	10
60–69	7	10

An example of grouped data that has different class intervals is presented in Table 1.4. This is a frequency table since it shows the frequency that data is represented in a specific class interval.

Table 1.4: Grouped data with different class intervals.

Daily Income (US$)	Frequency	Class Interval
0–9	15	10
10–19	18	10
20–29	17	10
30–49	35	20
50–79	20	30

1.3.2 Calculation of Class Interval for Ungrouped Data

In order to obtain a class interval for a given set of raw or ungrouped data, the following steps have to be taken:
1. Determine the number of classes you want to have
2. Subtract the lowest value from the highest value in the dataset
3. Divide the outcome under Step 2 by the number of classes you have in Step 1

The mathematical expression of these steps is below:

$$Class\ interval = \frac{HV - LV}{Number\ of\ classes}$$

where HV means highest value and LV means lowest value.

For instance, if a market survey is conducted on market prices for 20 non-durable goods with the following prices: US$11, US$5, US$3, US$14, US$1, US$16, US$2, US$12, US$2, US$4, US$3, US$9, US$4, US$8, US$7, US$5, US$8, US$6, US$10, US$15. The raw data indicate the lowest price is US$1 and the highest price is US$16. In addition, the survey expert decides to have 4 classes. The class interval is written as follows:

$$Class\ Interval = \frac{HV - LV}{Number\ of\ classes}$$

$$= \frac{16 - 1}{4} = \frac{15}{4} = 3.75$$

The value of the class interval is usually a whole number, but in this case its value is a decimal. Therefore, the solution to this problem is to round-off to the nearest whole number, which is 4. This implies that the raw data can be grouped into 4 as presented in Table 1.5.

Table 1.5: Class interval generated from ungrouped data.

Number	Frequency
1–4	7
5–8	6
9–12	4
13–16	3

1.3.3 Class Limits and Class Boundaries

Class limits are the actual values in the above-mentioned tables. In Table 1.5, **1** and **4** are the class limits of the first class. Class limits are divided into two categories: lower class limit and upper class limit. In Table 1.5 for the first class, **1** is the lower class limit while **4** is the upper class limit. Class limits are used in frequency tables.

The lower class boundary is obtained by adding the lower class limit of the upper class and the upper class limit of the lower class and dividing the answer by 2. The upper class boundary is obtained by adding the upper class limit of lower class and lower class limit of the upper class and dividing the answer by 2. Class boundaries reflect the true values of the class interval. In Table 1.6 to Table 1.6, the first line actually consists of values up to 4.5, which is the upper class boundary there. Class boundaries are also divided into lower and upper class boundaries. *Class boundaries* are not often observed in the frequency tables unless the data in the table is all integers. The class boundaries for the ungrouped data above is shown in Table 1.6.

Table 1.6: Class boundaries generated from ungrouped data.

Number	Frequency	Class Boundaries (Lower – Upper)
1–4	7	0.5–4.5
5–8	6	4.5–8.5
9–12	4	8.5–12.5
13–16	3	12.5–16.5

A class interval is the range of data in that particular class. The relationship between the class boundaries and the class interval is as follows:

Class Interval = Upper Class Boundary – Lower Class Boundary

The lower class boundary of one class is the same as the upper class boundary of the previous class for both integers and non-integers. The importance of class limits and class boundaries is highly recognized in the diagrammatical representation of statistical data.

1.4 Methods of Visualizing Data

There are many ways to plot data. In this section we will review some of the most common (shown in Figure 1.2).

1. **Time plot:** A *time plot* shows the history of several data items during the latest time horizon as lines. The time axis is always horizontal and directed to the right. The plot displays changes in a variable over time. It is also known as *line chart*.

2. **Pie chart:** A *pie chart* simply displays data as a percentage of a given total. This is the most appropriate way of visualizing quantities as percentages of a whole. The entire area of the pie is 100 percent of the quantity and the size of each slice is the percentage of the total represented by the category the slice denotes. Each percentage is determined by calculating the number of degrees for segments as a percentage of the 360 degrees in the circle.

3. **Bar chart:** *Bar charts* are mainly utilized to showcase categorical data without focus on the percentage. Its scale of measurement can be either nominal or ordinal. Bar charts can be displayed with the use of horizontal bars or vertical bars. A horizontal bar chart is more appropriate if someone intends to write the name of each category inside the rectangle that represents that category. A vertical bar chart is used to stress the height of different columns as measures of the quantity of the concerned variable.

4. **Frequency polygon:** A *frequency polygon* displays the same as a histogram, but without the rectangles. Instead, a point at the midpoint of each interval is used at a height corresponding to the frequency or a relative frequency of the category of the interval.

5. **Ogive:** An *ogive* refers to a cumulative frequency or cumulative relative-frequency graph. It begins at 0 and reaches a peak of 1.00 (for a relative-frequency ogive) or to the peak cumulative frequency of a variable. The point with height assigned to the cumulative frequency is situated at the right endpoint of each interval.

6. **Trellis plot:** A *trellis plot* is used for viewing complex or multi-variable data sets. It is called Trellis because they usually result in a rectangular array of plots, resembling a garden trellis. Trellis plots are based on conditioning the values taken on by one or more of the variables in a data set. Trellis plots are used to present complex data in a simplified form. When data analysis of split plot reports (involving two experiments with different sizes, for example) is being illustrated, a rectangular array may clarify what the experiment reveals. This is one example of how statistical inference may be more instructive than descriptive statistical methods alone.

7. **3D surface Plot:** A *3D surface plot* is a diagram of three-dimensional data and it show a functional relationship between a designated dependent variable (Y), and two independent variables (X and Z). The plot is a companion plot to the contour plot. These plots are useful in regression analysis for viewing the relationship among a dependent and two independent variables. Therefore, the 3D surface plot enables us to visually determine if multiple regression is appropriate.

Note: Charts can often be deceiving. This indicates the disadvantage of merely descriptive methods of analysis and the need for statistical inference. Exploring statistical tests makes the analysis more objective than eyeball analysis and less prone to deception if assumptions of random sampling and others are established.

Inflation in a country

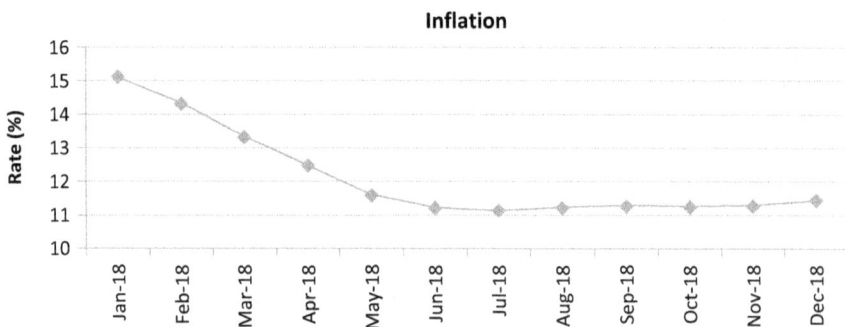

Figure 1.2(a): Time plot of inflation in Nigeria.
Source: NBS

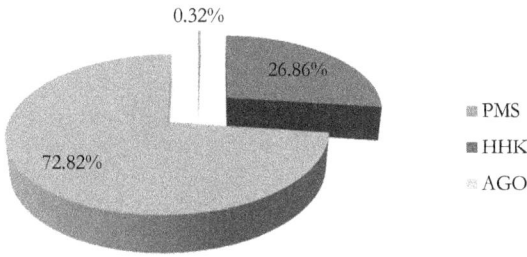

Figure 1.2(b): Pie chart of petroleum products distribution (in thousands of liters) in Nigeria.
Source: NNPC

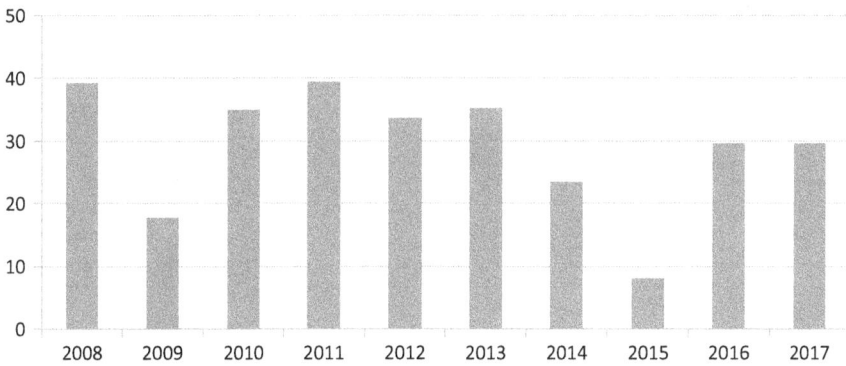

Figure 1.2(c): Bar chart of domestic crude oil processed (in billions of barrels) in Nigeria.
Source: NNPC

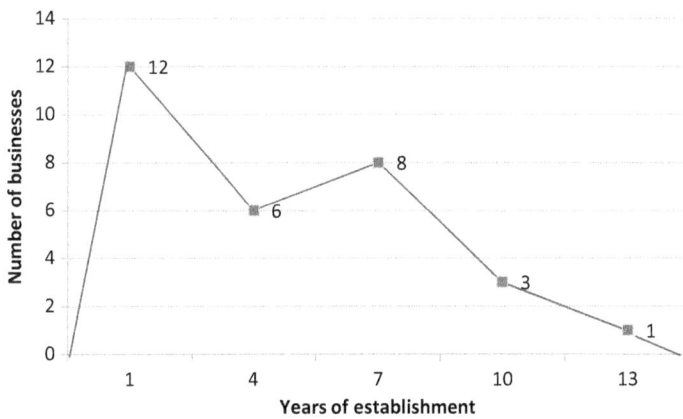

Figure 1.2(d): Polygon of number of businesses in years of establishment.

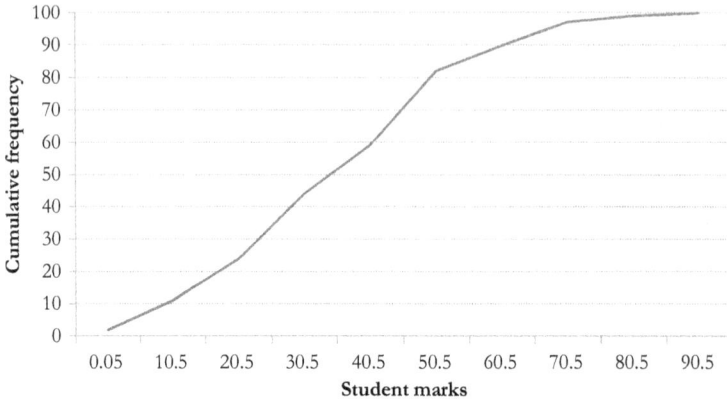

Figure 1.2(e): Ogive for the marks of students in an examination.

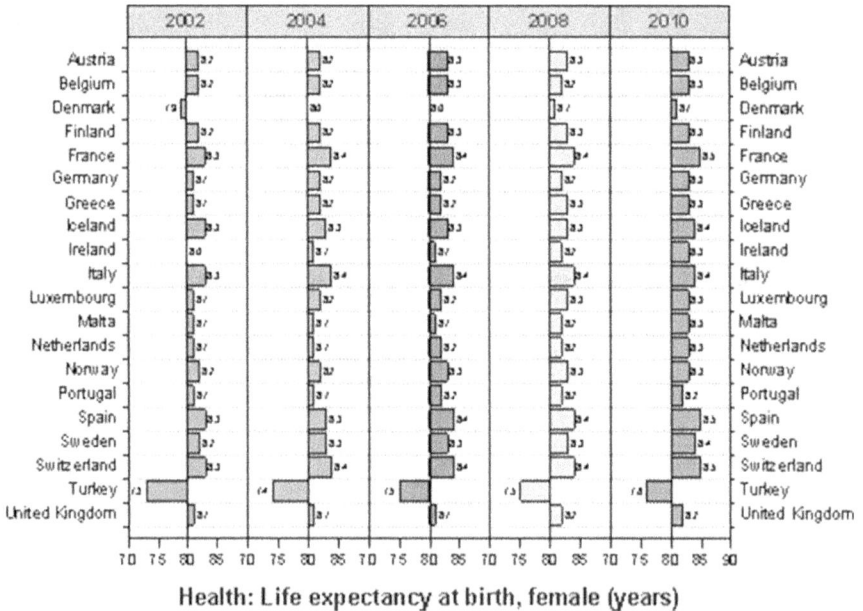

Health: Life expectancy at birth, female (years)

Figure 1.2(f): Trellis for life expectancy at birth, females (years).
Source: https://www.originlab.com/doc/Origin-Help/trellis

1.5 Exploring Data Analysis (EDA)

This is the name ascribed to a broad body of statistical and graphical methods. These approaches present patterns of observing data to determine relationships

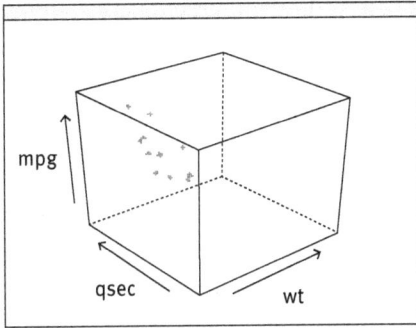

Figure 1.2(g): 3D surface plot of a cylinder.

and trends, identify outliers and influential observations, and quickly describe or summarize datasets. This analysis was initiated by the work of John W. Tukey, "Exploring Data Analysis," in 1977, and has made significant strides in the past five years as software solutions in large data analysis (Big Data) and in business intelligence (the reporting of characteristics of data in meaningful ways) has improved dramatically.

1.5.1 Stem-and-Leaf Displays

A *stem-and-leaf display* presents a quick way of observing a dataset. It has some attributes of a histogram but prevents the loss of information that a histogram has by organizing data into intervals. This type of display adopts the tallying principles as well as the use of the decimal-based number system. The stem refers to the number without its rightmost digit (the leaf). The stem is written to the left of a vertical line separating the stem from the leaf.

For example, if we have the numbers 201, 202, 203, and 205, the stem-and-leaf display is:

20|1235

1.5.2 Box Plots

A *box plot* is also known as a *box-whisker plot* and visually presents a dataset by determining its central tendency, spread, skewness, and the presence of outliers. Spacing between parts of the box indicate the degree of dispersion and skewness in the data as well as identifying probable outliers. The box plot consists of the five summary measures of the distribution of data: median, lower quartile, upper quartile, the smallest observation, and the largest observation. It has two fundamental assumptions. The first assumption is that the hinges of the box plot are the quartiles of the dataset, and the median is a line inside this box. The second assumption is that the whiskers are constructed by extending a line from the upper quartile to the largest observation, and from the lower quartile to the smallest observation, under the condition that the largest and smallest observations lie within a distance of 1.5 times the interquartile range from the appropriate hinge (quartile). Observations that exceed this distance are regarded as suspected outliers. Figure 1.3 shows a typical example of a Box-Whisker plot.

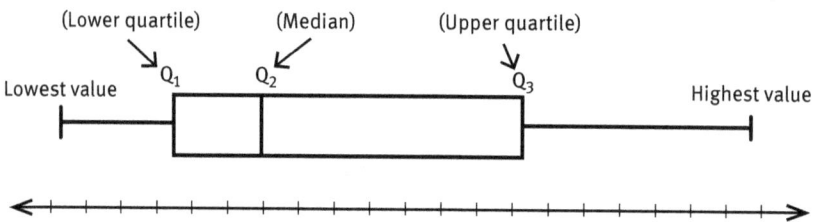

Figure 1.3: Box-whisker plot.

1.5.2.1 Importance of Box Plots
a. They are used to identify the location of a dataset using the median.
b. They are explored to determine the spread of the data through the length of the box, the interquartile range, and the length of the whiskers.
c. They provide a means of identifying possible skewness of the distribution of the dataset.
d. They detect suspected outliers in the dataset.
e. They are useful for comparing two or more datasets by drawing box plots for each dataset and displaying the box plots on the same scale.

Exercises for Chapter 1

1. Explain why statistics are necessary.
2. What is the difference between a quantitative variable and a qualitative variable?
3. List the four scale of measurements and discuss each with relevant examples.
4. The total liters of petroleum product importation and consumption in a country in the second quarter of 2018 is summarized in the table below.

Months	PMS	AGO	HHK	ATK
April	1,781,949,720	309,988,876	0	14,700,325
May	1,670,390,264	387,434,670	0	64,186,515
June	1,340,777,749	408,153,067	43,790,592	121,501,304
Total	4,793,117,733	1,105,576,613	43,790,592	200,388,144

(a) Represent the liters of petroleum product importation and consumption in a bar chart.

(b) Represent the total liters of petroleum product importation and consumption by products in a pie chart.

2 Introduction to R Software

R is a programming language designed for statistical analysis and graphics. It is based on S-plus which was developed by Ross Ihaka and Robert Gentleman from the University of Auckland, New Zealand. R can be used to open multiple datasets. R is open-source software that can be downloaded at http://cran.r-project.org/. There are other statistical packages—SPSS, SAS, and Stata—but they are not open source. In addition, there are large R user groups online that can provide real-time answers to questions and that also contribute packages to R. Packages increase the functions that are available for use in R, thus expanding the users' capabilities. The R Development Core Team is responsible for maintaining the source code of R. This chapter covers a lot of ground, but it is impossible to document R in a chapter and so we provide explanations and examples of the significant statistical functions. The R code is provided in many examples throughout this book.

Why turn to R?
R software provides the following advantages:
1. R is free (meaning it is open-source software).
2. Any type of data analysis can be executed in R.
3. R includes advanced statistical procedures not yet present in other packages.
4. The most comprehensive and powerful feature for visualizing complex data is available in R.
5. Importing data from a wide variety of sources can be easily done with R.
6. R is able to access data directly from web pages, social media sites, and a wide range of online data services.
7. R software provides an easy and straightforward platform for programming new statistical techniques.
8. It is easy to integrate applications written in other programming languages (such as C++, Java, Python, PHP, Pentaho, SAS, and SPSS) into R.
9. R can operate on any operating system such as Windows, Unix, and Mac OSX. It can also be installed on an iPhone. It is possible to use R on an Android phone (see https://selbydavid.com/2017/12/29/r-android/).
10. R offers a variety of graphical user interfaces (GUIs) if you are not interested in learning a new language.

Packages in R
Packages refer to the collections of R functions, data, and compiled code in a well-defined format.

https://doi.org/10.1515/9781547401475-002

2.1 How to Download and Install R

Many operating systems such as Windows, Macintosh OSX, and Linux can connect with R. R is a free download and it can be accessed from many different servers around the world. All available download links can be found on http://www.r-project.org/index.html by viewing the "Get Started" section on the front page and clicking on "download R." These mirrors are linked with the word "CRAN" which stands for the **C**omprehensive **R** **A**rchive **N**etwork. The CRAN provides the most recent version of R.

1. By choosing the mirror at the top of your computer screen, a list of the versions of R for each operating system is displayed.
2. Click the R version that is compatible with your operating system.
3. If you decide to have the base version, just click the "download" link that is displayed on the screen.

Installation of R

1. Under the "download" link, there is another link that provides instructions on how to install R. These might be useful if you encounter problems during the installation process.
2. To install R, you double-click on the executable file and follow the instruction on the screen. The default settings are perfect. Figure 2.1 is the first screen that would display if you are installing R on a Windows system.

2.2 Using R for Descriptive Statistical Analysis and Plots

R can be utilized to develop any statistical analysis. Different types of graphs can be constructed using R statistical software. A range of standard statistical plots such as scatterplots, boxplots, histograms, bar plots, pie charts, and basic 3D plots is provided in R. These basic types of plots can be generated using a single function call in R. The R graphics are able to add several graphical elements together to produce the final outcome. Apart from the conventional statistical plots, R also generates Trellis plots through the package "lattice." Special purpose plots like lines, text, rectangles, and polygons are also available in the R software. Owing to its generality and feasibility, production of graphical images that are above the normal statistical graphics is possible. This also allows R to generate figures that can visualize important concepts or useful insights. The structure of R graphics has four distinct levels, namely: graphics packages; graphics systems; a graphics engine; and graphics device packages (the full application will be discussed throughout the rest of this book).

Steps to install R

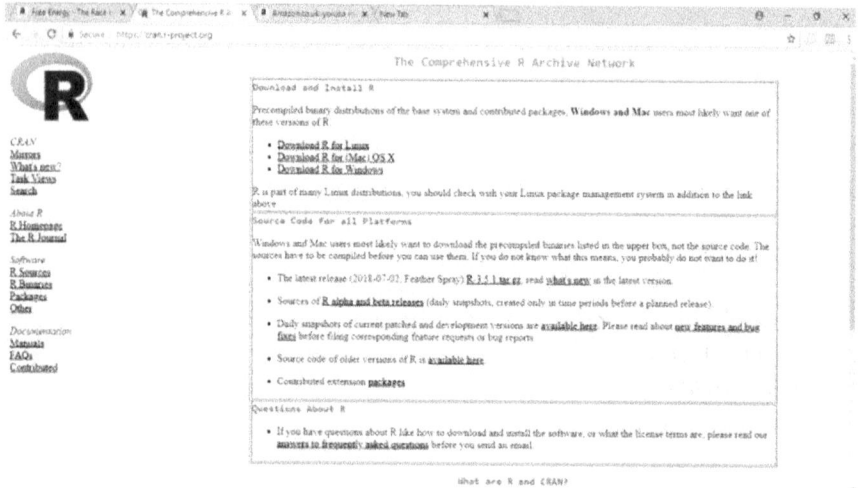

Click the appropriate operating system based on your computer. For example, "Download R for Windows."

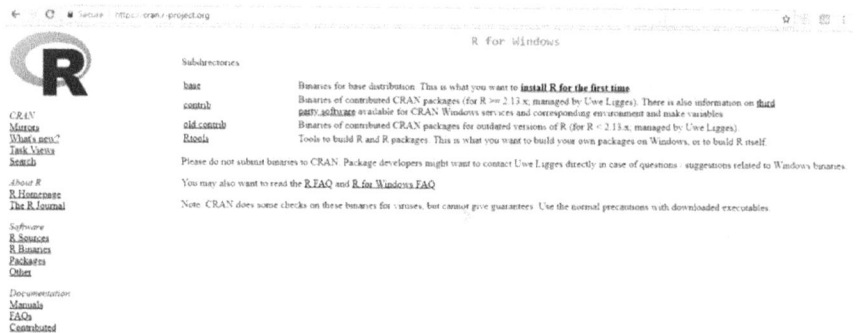

Click on "R Sources" in the menu on the left side of the page under "Software"to get information on the latest version of R.

Figure 2.1: Installation processes in R.

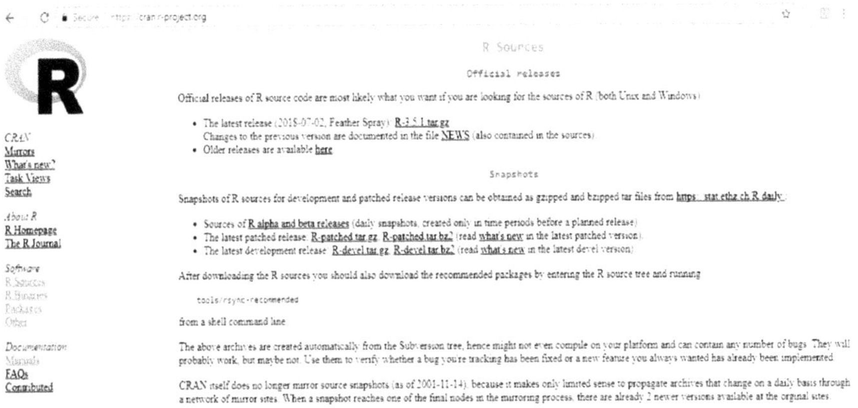

Click on "Packages,"also on the left side menu under "Software," to browse all available packages in R software.

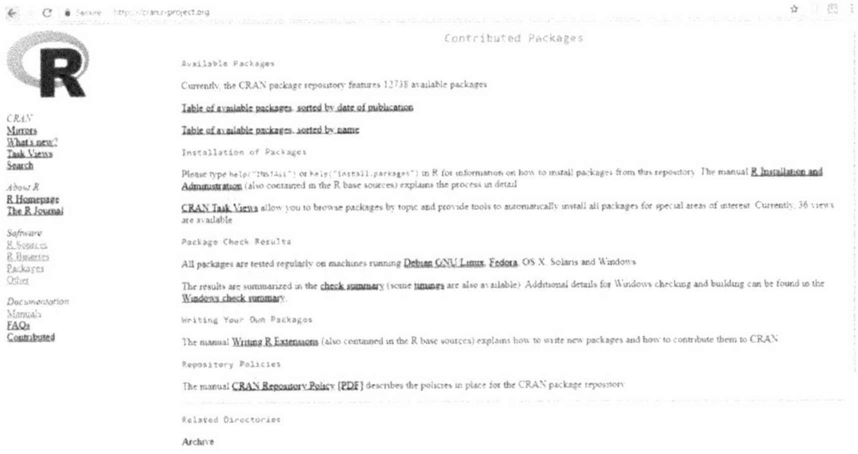

Click on "Manuals" under "Documentation" on the left-side menu to browse the R user guides.

Figure 2.1 (continued)

Figure 2.1 (continued)

2.3 Basics of R

2.3.1 R Is Vectorized

Some of the functions in R are *vectorized*. This indicates that the functions operate on the elements of a vector by acting on each element one after the other
without undergoing looping processes. Vectorization allows the user to write
code in R that is easy to read, efficient, precise, and concise. The following examples demonstrate vectorization in R:

Example 2.1:
The R language allows multiplying (or dividing) a vector by a constant value. A simple example is:

```
x<-c(2, 4, 6, 8, 10)
x*3
[1] 6 12 18 24 30
x/3
[1] 0.6666667 1.3333333 2.0000000 2.6666667 3.3333333
```

Example 2.2:
The R language allows addition (or subtraction) of vectors.

```
x<-c(2, 4, 6, 8, 10)
y<-c(1, 2, 3, 4, 5)
x+y
```

```
[1] 3 6 9 12 15
x-y
[1] 1 2 3 4 5
```

Example 2.3:
R allows for the multiplication of vectors.

```
x<-c(2, 4, 6, 8, 10)
y<-c(1, 2, 3, 4, 5)
x*y
[1] 2 8 18 32 50
```

Example 2.4:
R allows for the use of logical operations.

```
x<-c(2, 4, 6, 8, 10)
a<-(x>=4)

a
[1] FALSE TRUE TRUE TRUE TRUE
```

Example 2.5:
R allows for the use of matrix operations. Given that matrix a and matrix b are 2 by 2 matrices In R, the matrices shown below are coded as follows:

```
a <- matrix(1:4, 2, 2)
b <- matrix(0:3, 2, 2)
a
      [,1] [,2]
[1,]    1    3
[2,]    2    4
b
      [,1] [,2]
[1,]    0    2
[2,]    1    3
```

In our R example below, we multiply the elements by position in each matrix by the corresponding element in the other matrix. The code above tells R to create a matrix using the numbers from 1 to 4 and another from 0 to 3 and the default is that the elements are placed down rows and then put in the next column. In this case the matrices are 2 x 2.

```
# the product of matrix a and matrix b
ab<-a*b
ab
    [,1] [,2]
[1,]  0  6
[2,]  2  12
```

The result of the matrix is saved in a matrix ab.

2.3.2 R Data Types

There are different types of data in R: scalars, vectors, matrices, data frames, lists, factors, coercions, and date and time.

2.3.2.1 Scalars

A *scalar* is an atomic quantity that can hold only one value at a time. Scalars are the most basic data types that can be used to construct more complex ones. Scalars in R can be numerical, logical, and character(string). The following points are examples of types of scalars in R.

- **Numerical**
```
p<-10     # assigning the value 10 to p
class(p)
[1] "numeric"

q<-12     # assigning the value 12 to q
class(q)
[1] "numeric"

#Addition of two numerical values is also numeric
class(p+q)
[1] "numeric"
```

- **Logical**
A logical statement is a statement which can either be true or false. Let's assume the values p = 10 and q = 12.

The first segment of the R code checks if the value of p is less than q (i.e., is 10<12?). Definitely, the answer is yes.

```
t<-  p<q  # Is p less than q?
t
[1] TRUE
```

```
class(t)
[1] "logical"
```

The result here is TRUE because the statement is correct and the value of t is logical.

```
s<- p>q  # Is p greater than q?
s
[1] FALSE
class(s)
[1] "logical"
```

The result here is FALSE because $10 \not> 12$
and the value of t is logical.

```
class(NA)
[1] "logical"
```

So in addition, the NA is classified as logical.

- **Character (string)**
```
m<-"10"
n<-"12"
```

To confirm that the values of m and n are string characters
```
m
  [1] "10"
n
  [1] "12"
  m+n     # this is not the same as p and q in the earlier example
  Error in m +n : non-numeric argument to binary operator

  class(m)
  [1] "character"

  class(n)
  [1] "character"

  class(as.numeric(m))
  [1] "numeric"
```

2.3.2.2 Vectors
A *vector* is a sequence of data elements of the same basic type.

```
d<-c(1, 2, 3, 4, 5, 6) # Numeric vector
class(d)
[1] "numeric"
```

```
e<-c("one", "two", "three", "four", "five", "six") # Character vector
class(e)
[1] "character"

f<-c(TRUE, FALSE, TRUE, TRUE, FALSE, TRUE)   #Logical vector
class(f)
[1] "logical"

d;e;f
[1] 1 2 3 4 5 6
[1] "one"   "two"   "three" "four" "five" "six"
[1] TRUE FALSE TRUE TRUE FALSE TRUE
```

Braces [ ] are used to reference elements in a vector. Let's demonstrate the uses of the brace [] symbol by writing a short R function to reference some points of a coordinate and display our results using the print command.

```
points<-c(c(-2,5),c(10,-22),c(12, 28))
for(i in seq(1,length(points),2)){
print(points [i])
print(points [i+1])
}

[1] -2
[1] 5
[1] 10
[1] -22
[1] 12
[1] 28
```

2.3.2.3 Matrices
A *matrix* is a collection of data elements arranged in a two-dimensional rectangular form. In the same manner as a vector, the components in a matrix must be of the same basic type. An example of a matrix with 2 rows and 4 columns is created below.

```
# fill the matrix with elements arranged by column in 2 rows and 4 columns
mat<-matrix(1:8, nrow=2, ncol=4, byrow= FALSE)
mat
     [,1] [,2] [,3] [,4]
[1,]  1   3   5   7
[2,]  2   4   6   8
```

Alternatively, it is possible to have the elements of the matrix arranged by rows.

```
# fill the matrix with elements arranged by row in 2 rows and 4 columns
mat<-matrix(1:8, nrow=2, ncol=4, byrow= TRUE)
mat
     [,1] [,2] [,3] [,4]
[1,]  1   2   3   4
[2,]    5  6  7  8
```

Braces [ ] can be used to reference elements in a matrix; this is similar to referencing elements in vectors.

```
mat[2,3]  # refers to the element in the second row and third column
[1] 7
mat[,4]  # refers to all elements in the fourth column
[1] 4 8
mat[1,]  # refers to all elements in the first row
[1] 1 2 3 4
```

Table 2.1 shows the basic matrix operations and their respective meanings.

Table 2.1: Basic functions and their meanings.

Function	Meaning
t(x)	Transposition of x
diag(x)	Diagonal elements of x
%*%	Matrix multiplication
solve (a,b)	Solves a %*% x = b for x[a]
rowsum(x)	Sum of rows for a matrix-like object; rowSums(x) is a faster version
rowMeans(x)	Fast version of row means
colMeans(x)	Fast version of column means

[a] This generic function solves the equation a %*% x = b for x, where b can be either a vector or a matrix.

2.3.2.4 Data Frames

A *data frame* is more general than a matrix in the sense that different columns can have different basic data types. A data frame is one of the most common data types presented for data analysis. Let's consider combining the vectors of numerical, character, and logical vectors to form a data frame. Data frames have a special attribute called *row names*. Data frames are created through **read.table()** or **read.csv()**.

```
d<-c(1, 2, 3, 4,  5, 6)
e<-c("one", "two", "three", "four", "five", "six")
f<-c(TRUE, FALSE, TRUE, TRUE, FALSE, TRUE)
data<-data.frame(d, e, f)
names(data)<-c("ID", "words", "state")
data
    ID words state
  1 1   one   TRUE
  2 2   two   FALSE
  3 3   three TRUE
  4 4   four  TRUE
  5 5   five  FALSE
  6 6   six   TRUE
```

In addition, components from data frames can be extracted. This is similar to the extraction of components in matrices, but assigning a name to each column makes it more flexible. For example:

```
data$ID
[1] 1 2 3 4 5 6

data[1:2,]
    ID words state
1 1   one TRUE
2 2   two FALSE

data[,3]
[1] TRUE FALSE TRUE TRUE FALSE TRUE
```

2.3.2.5 Lists

A *list* is a generic vector containing other objects. There are two characteristics of this type of vector: (1) a list can contain elements of different classes, and (2) each element of a list can have a name. There are no restrictions on data types or length of the components. It is easier to work with lists that have named components.

```
Consider this list which contains a vector, matrix, and data frame.
lst<-list(vector=d, matrix=mat, frame=data, count=10)
    lst
    $vector
    [1] 1 2 3 4 5 6

    $matrix
         [,1] [,2] [,3] [,4]
    [1,]  1    3    5    7
    [2,]  2    4    6    8

    $frame
    ID words state
    1 1 one    TRUE
    2 2 two    FALSE
    3 3 three  TRUE
    4 4 four   TRUE
    5 5 five   FALSE
    6 6 six    TRUE

    $count
    [1] 10
```

2.3.2.6 Factors

Factors are used to represent categorical data and can be ordered or unordered. We can regard a factor as a numerical vector when each of the integers contains a label. It is more appropriate to use factors with labels rather than integers because factors are self-describing (e.g., assign a variable with the labels "Male" and "Female" rather than assigning the values 1 and 2 to them).

```
y<-factor(c("yes", "yes", "yes", "no", "no"))
y
  [1] yes yes yes no no
  Levels: no yes

  table(y) # tabulate the outcome
  y
    no yes
    2   3
```

2.3.2.7 Coercions

Coercion occurs when different data types are mixed in a vector. Since a vector can only be composed of one data type, R will coerce every element in the vector to the same class.

```
co<-c(50, "big") # character
co
[1] "50" "big"
```

The numerical value of 50 is converted to a character.

```
co<-c(TRUE, 2) # numeric
co
[1] 1 2
```

The logical value TRUE is converted to numerical value 1.

```
co<-c(FALSE, 2) # numeric
co
[1] 0 2
```

The logical value FALSE is converted to numerical value 0.

2.3.2.8 Date and Time

The date and time can be presented in the "POSIXct" class and "POSIXlt" class. The class "POSIXct" represents the (signed) number of seconds since the beginning of 1970 as a numeric vector. The class "POSIXlt" is a named list of vectors representing:

min   0–59 minutes
hour   0–23 hours
yday   0–365 day of the year

A date and time that is represented by by the POSIXct class is a very large integer. The POSIXlt class is a list that stores a lot of useful metadata (i.e., data that describes other data). The function **as.Date()** is used to convert strings to dates.

```
x <- as.Date("2018-03-21")
x
[1] "2018-03-21"

# use as.Date( ) to convert strings to dates
dates <- as.Date(c("2018-03-21", "2018-10-04"))
```

```
# number of days between 03/21/2018 and 10/04/2018
days <- dates[2] - dates[1]
days
 Time difference of 197 days
```

In addition, **Sys.Date()** returns today's date, **date()** returns the current date and time, and **strptime ()** is used to write times in a different format than characters.

Table 2.2 shows symbols that can be used with the **format()** function to print dates:

Table 2.2: Some symbols with their meanings.

Symbol	Meaning	Example
%d	day as a number (0–31)	31-Jan
%a	abbreviated weekday	Fri
%A	unabbreviated weekday	Friday
%m	month	00–12
%b	abbreviated month	Mar
%B	unabbreviated month	March
%y	year without century	00–99
%Y	year with century	2018

```
To print today's date using the Sys.Date( ) function:
today <- Sys.Date() # print today's date
format(today, format="%B %d %Y")
[1] "October 04 2018"

x <- Sys.time( )
x
[1] "2018-10-04 11:35:15 WAT"

p<-as.POSIXlt (x)
names (unclass(p))  # unclass (p) is a list object
[1] "sec" "min" "hour" "mday" "mon" "year" "wday" "yday"
[9] "isdst" "zone" "gmtoff"
p$sec
[1] 15.44309
p$wday
[1] 4
p$mday
```

```
[1] 4
timeString<-"October 4, 2018 13:58"
h<-strptime(timeString, "%B %d, %Y %H: %M")
class(h)
[1] "POSIXlt" "POSIXt"
h
[1] "2018-10-04 13:58:00 WAT"
```

2.3.3 Missing Values

Missing values are represented by NA (not available) and should be capitalized. Undefined values are represented by NaN (not a number). NaN is also Na but the converse of this is not true. For example, we have:

```
mv<-c(1, 2, 3, NA, 5)
is.na(mv) # returns TRUE if mv is missing
[1] FALSE FALSE FALSE TRUE FALSE
```

In addition, we can exclude missing values from an analysis:

```
mv<-c(1, 2, 3, NA, 5)
mean(mv) # returns NA
[1] NA
mean(mv, na.rm=T) # returns average of 2.75 by excluding the missing value
[1] 2.75
```

The function **na.omit()** returns the object with listwise deletion of missing values.

```
# create new dataset without missing data
d<-c(1, 2, 3, NA, 5, 6)
e<-c("one", "two", "three", "four", "five", "six")
f<-c(TRUE, FALSE, TRUE, TRUE, FALSE, TRUE)
data1<-data.frame(d, e, f)
data1
    d e f
1 1  one   TRUE
2 2  two   FALSE
3 3  three TRUE
4 NA four  TRUE
5 5  five  FALSE
6 6  six   TRUE
```

```
newdata <- na.omit(data1)
newdata  # delete all entries for row 4 since it contains the missing value
    d e f
1 1  one   TRUE
2 2  two   FALSE
3 3  three TRUE
5 5  five  FALSE
```

2.3.4 Data Creation

In this subsection, we discuss how datasets can be generated in different ways, examples are used to demonstrate each of the R commands.

c(...)—A generic function that combines arguments with the aim of forming a vector; when *recursive = TRUE* in the argument, then it descends through lists combining all elements into one vector.

from: to— generates a sequence. The colon symbol ":" has operator priority.

```
x<-1:4+1
x[1] 2 3 4 5

y<-10:14+6
y
[1] 16 17 18 19 20
```

seq(from, to)— Generates a sequence : = specifies increment, length = specifies desired length.

    seq( ) function generates a sequence of numbers.

    seq(from = 1, to = 1, by = ((to - from)/(length.out - 1)), length.out = NULL, along.with = NULL, ...)

    from, to: beginning and end numbers of the sequence

    by: step, increment (default is 1)

    length.out: length of the sequence

```
# generate a sequence from -2 to 10, step 2
s<-seq(-2, 10, 2)
s
 [1] -2 0 2 4 6 8 10

# generate a sequence with 15 evenly distributed from -2 to 10
 s1<-seq(-2, 10, length.out=15)
```

```
s1
[1] -2.0000000 -1.1428571 -0.2857143 0.5714286 1.4285714 2.2857143
[7] 3.1428571 4.0000000 4.8571429 5.7142857 6.5714286 7.4285714
[13] 8.2857143 9.1428571 10.0000000
```

rep(x, times) : replicate x times; use = to repeat each element of x a specified number of times;

```
rep(c(1, 2, 3, 4, 5), 2) #repeat 1 2 3 4 5 twice
[1] 1 2 3 4 5 1 2 3 4 5

rep(c(1, 2, 3, 4, 5), each = 2) # repeat each of 1 2 3 4 5 twice
[1] 1 1 2 2 3 3 4 4 5 5

replicate(3, c(1, 3, 6)) # repeat vector 1 3 6 thrice

     [,1] [,2] [,3]
[1,]  1    1    1
[2,]  3    3    3
[3,]  6    6    6
```

Note that *replicate* is used for the repetition of vectors

array(x, dim =):——Used to create an array with data x and then specify dimensions (e.g., dim=c(3, 4, 2); the elements of x recycle if c is not long enough). We can give names to the rows, columns, and matrices in the array by using the dimnames parameter as shown in the following R code:

```
# Create two vectors of different lengths.
vector1 <- c(1,2,3)
vector2 <- c(4,5,6,7,8,9)
column.names <- c("Col1","Col2","Col3")
row.names <- c("Row1","Row2","Row3")
matrix.names <- c("Matrix1","Matrix2")
arr <- array(c(vector1,vector2),dim = c(3,3,2),dimnames = list(row.names,column.
names,
  matrix.names))
arr

,, Matrix1

   Col1 Col2 Col3
```

```
Row1  1  4  7
Row2  2  5  8
Row3  3  6  9

,,Matrix2

   Col1 Col2 Col3
Row1  1  4  7
Row2  2  5  8
Row3  3  6  9
```

To generate factors by specifying the pattern of their levels:

> *gl(n, k, length=n*k, labels = 1:n)*

where *k* is the number of levels, and *n* is the number of replications; *length* is an integer giving the length of the result; *labels* is an optional vector of labels for the resulting factor levels.

Create a data frame from all combinations of the supplied vectors or factors:

> *expand.grid(. . ., KEEP.OUT.ATTRS = TRUE, stringsAsFactors = TRUE)*

where the argument ". . ." denotes the vectors, factors or a list containing these

> *KEEP.OUT.ATTRS* is a logical argument indicating the "out.attrs" attribute
> *stringsAsFactors* is a logical argument specifying if character vectors are converted to factors.

```
expand.grid(height = seq(1.5, 2.2, 0.05), weight = seq(50, 70, 5),sex = c
("Male","Female"))

   height weight  sex
1    1.50    50  Male
2    1.55    50  Male
3    1.60    50  Male
4    1.65    50  Male
5    1.70    50  Male
6    1.75    50  Male
7    1.80    50  Male
8    1.85    50  Male
9    1.90    50  Male
10   1.95    50  Male
11   2.00    50  Male
12   2.05    50  Male
13   2.10    50  Male
```

```
14  2.15  50  Male
15  2.20  50  Male
16  1.50  55  Male
17  1.55  55  Male
18  1.60  55  Male
19  1.65  55  Male
20  1.70  55  Male
21  1.75  55  Male
22  1.80  55  Male
23  1.85  55  Male
24  1.90  55  Male
25  1.95  55  Male
26  2.00  55  Male
27  2.05  55  Male
28  2.10  55  Male
29  2.15  55  Male
30  2.20  55  Male
31  1.50  60  Male
32  1.55  60  Male
33  1.60  60  Male
34  1.65  60  Male
35  1.70  60  Male
36  1.75  60  Male
37  1.80  60  Male
38  1.85  60  Male
39  1.90  60  Male
40  1.95  60  Male
41  2.00  60  Male
42  2.05  60  Male
43  2.10  60  Male
44  2.15  60  Male
45  2.20  60  Male
46  1.50  65  Male
47  1.55  65  Male
48  1.60  65  Male
49  1.65  65  Male
50  1.70  65  Male
51  1.75  65  Male
52  1.80  65  Male
53  1.85  65  Male
54  1.90  65  Male
55  1.95  65  Male
56  2.00  65  Male
```

57	2.05	65	Male
58	2.10	65	Male
59	2.15	65	Male
60	2.20	65	Male
61	1.50	70	Male
62	1.55	70	Male
63	1.60	70	Male
64	1.65	70	Male
65	1.70	70	Male
66	1.75	70	Male
67	1.80	70	Male
68	1.85	70	Male
69	1.90	70	Male
70	1.95	70	Male
71	2.00	70	Male
72	2.05	70	Male
73	2.10	70	Male
74	2.15	70	Male
75	2.20	70	Male
76	1.50	50	Female
77	1.55	50	Female
78	1.60	50	Female
79	1.65	50	Female
80	1.70	50	Female
81	1.75	50	Female
82	1.80	50	Female
83	1.85	50	Female
84	1.90	50	Female
85	1.95	50	Female
86	2.00	50	Female
87	2.05	50	Female
88	2.10	50	Female
89	2.15	50	Female
90	2.20	50	Female
91	1.50	55	Female
92	1.55	55	Female
93	1.60	55	Female
94	1.65	55	Female
95	1.70	55	Female
96	1.75	55	Female
97	1.80	55	Female
98	1.85	55	Female
99	1.90	55	Female

```
100  1.95  55  Female
101  2.00  55  Female
102  2.05  55  Female
103  2.10  55  Female
104  2.15  55  Female
105  2.20  55  Female
106  1.50  60  Female
107  1.55  60  Female
108  1.60  60  Female
109  1.65  60  Female
110  1.70  60  Female
111  1.75  60  Female
112  1.80  60  Female
113  1.85  60  Female
114  1.90  60  Female
115  1.95  60  Female
116  2.00  60  Female
117  2.05  60  Female
118  2.10  60  Female
119  2.15  60  Female
120  2.20  60  Female
121  1.50  65  Female
122  1.55  65  Female
123  1.60  65  Female
124  1.65  65  Female
125  1.70  65  Female
126  1.75  65  Female
127  1.80  65  Female
128  1.85  65  Female
129  1.90  65  Female
130  1.95  65  Female
131  2.00  65  Female
132  2.05  65  Female
133  2.10  65  Female
134  2.15  65  Female
135  2.20  65  Female
136  1.50  70  Female
137  1.55  70  Female
138  1.60  70  Female
139  1.65  70  Female
140  1.70  70  Female
141  1.75  70  Female
142  1.80  70  Female
```

```
143  1.85  70  Female
144  1.90  70  Female
145  1.95  70  Female
146  2.00  70  Female
147  2.05  70  Female
148  2.10  70  Female
149  2.15  70  Female
150  2.20  70  Female
```

The function **rbind** is used to combine arguments by rows for matrices, data frames, and other data types. When you have different vectors of datasets and you would like to use them for comparison, you may join them together to form a data frame.

```
r1<-c(1:10)
r2<-c(1, 3, 5, 7, 9, 11,13,15, 17, 19)
rbind(r1,r2)
   [,1] [,2] [,3] [,4] [,5] [,6] [,7] [,8] [,9] [,10]
r1  1  2  3  4  5  6  7  8  9  10
r2  1  3  5  7  9  11  13 15 17  19
```

The function **cbind** is used to combine arguments by columns for matrices, data frames, and data types. This function is similar to rbind but the main difference is that the vectors of datasets are combined by column and it is useful in forming a data frame for the purpose of modeling and analysis.

```
c1<-c(1:10)
c2<-c(1,3,5,7,9, 11,13,15, 17, 19)
cbind(c1,c2)
     c1 c2
 [1,] 1 1
 [2,] 2 3
 [3,] 3 5
 [4,] 4 7
 [5,] 5 9
 [6,] 6 11
 [7,] 7 13
 [8,] 8 15
 [9,] 9 17
[10,] 10 19
```

2.3.5 Data Type Conversion

A data type can be changed in R. Adding a character string to a numeric vector converts all the elements in the vector to character type. Data type conversion is necessary in the following situations: when two or more operands of different types appear in an expression; when arguments that do not conform exactly to the parameters declared in a function prototype are passed to a function; and when the data type of an operand is deliberately converted by the cast operator.

The following functions can be used in the conversion of a data type:

is.numeric(), is.character(), is.vector(), is.matrix(), is.data.frame(), as.numeric(), as.character(), as.vector(), as.matrix(), as.data.frame().

2.3.6 Variable Information

Here is a list of R functions on variables and their respective uses.

- *is.na(x), is.null(x), is.array(x), is.data.frame(x), is.numeric(x), is complex(x), is character(x)*—test for types; for a complete list, use method (is)
- *length(x)*—number of elements in x
- *dim(x)*—retrieve or set the dimension of an object; *dim<-c(3,2)*
- *dimnames(x)*—retrieve or set the name of the dimension of an object; x <- matrix(0:3, nrow=2, dimnames=list(c("R1", "R2"), c("C1", "C2"))) x

    C1 C2

    R1 0 2
- R2 1 3 *nrow(x)*—number of rows; NROW(x) is the same but treats a vector as a one-row matrix
- *class(x)*—get or set the class attribute of x; class(x)<-"myclass"
- *(unclass(x))*—remove the class attribute of x
- *attr(x,which)*—get or set the attribute which of x
- *attribute(obj)*—get or set the list of attributes obj

2.4 Basic Operations in R

For a list of numbers in memory, basic operations can be performed. These operations can be used to quickly perform a large number of calculations with a single command. It is obvious that if an operation is performed on more than one vector it is often necessary that the vectors all contain the same number of entries. In this section, we shall focus on the subsetting, control structures, built-in functions, user-written functions, importing, reporting and writing data.

2.4.1 Subsetting

Subsetting is an operation that can be used to extract subsets of R objects. For example, the symbol [always returns an object of the same class as the original object. With one exception, it can be used to select more than one element. The symbol [[is used to extract elements of a list or a data frame. $ is used to extract elements of a list or data frame by name. The semantics of $ are similar to [[.

In addition, an element of a vector v is assigned an index by its position in the sequence, starting with 1. The basic function for subsetting is []. v[1] is the first element, v[length(v)] is the last. The subsetting function takes input in many forms.

Example 2.6:

```
sub<-c("a", "b", "c", "d", "e", "f")
sub[4]
[1] "d"

sub[2:5]
[1] "b" "c" "d" "e"

sub>"c"
[1] FALSE FALSE FALSE TRUE TRUE TRUE

sub[sub>"c"]
[1] "d" "e" "f"
```

Example 2.7:

```
sub1<-matrix(1:9,3,3)
sub1
     [,1] [,2] [,3]
[1,]   1    4    7
[2,]   2    5    8
[3,]   3    6    9
sub1[1,3]
[1] 7
sub1[2,]    # entire elements in the second row
[1] 2 5 8
sub1[,1]    # entire elements in the first column
[1] 1 2 3
```

Example 2.8:

```
sub<-c("a", "b", "c", "d", "e", "f")
v <- c(1, 3, 6)
sub[v]
[1] "a" "c" "e"
v[1:3]
[1] "a" "b" "c"

sub<-c("a", "b", "c", "d", "e", "f")
v <- c(1, 3, 6)
sub[v]
[1] "a" "c" "f"
sub[1:3]
[1] "a" "b" "c"
```

Example 2.9:

```
sub1<-matrix(1:9,3,3)
sub1[1,3]
[1] 7
sub1[1,2, drop = FALSE] # return as a matrix of 1 by 1
   [,1]
[1,]  4
sub1[1,2, drop = TRUE] # return as a single element
[1] 4
sub1[1,, drop = TRUE] # return as a vector
[1] 1 4 7
```

Example 2.10:

```
y<-list(w = 1:3, x = 0.5, z ="gender")
y[1]

$w
[1] 1 2 3

y$w
[1] 1 2 3
```

```
y$x
[1] 0.5

y$z
[1] "gender"
```

2.4.2 Control Structures

Control structures allow you to control the flow of execution of a script, typi-
cally inside of a function. These control structures can not be used while
working with R interactively. We shall consider some common control structures in this section. These include: *conditional (if-else), for-loop, repeat-loop,
while-loop.*

2.4.2.1 Conditional
The conditional if … else statement is used for decision making in programming.

```
if (condition) {
  # do something
} else {
  # do something else
}
```

Example 2.11:

```
x <- 1:20
if (sample(x, 1) <= 10) {
  print("x is less than 10")
  } else {
  print("x is greater than 10")
  }
[1] "x is greater than 10"
```

2.4.2.2 For Loop
Loops are used in programming to repeat a specific block of code. A *for loop*
works on an iterable variable and assigns successive values until the end of
a sequence.

Example 2.12:

```
for (i in 1:10) {
  print(i*2)
  }
```

```
[1] 2
[1] 4
[1] 6
[1] 8
[1] 10
[1] 12
[1] 14
[1] 16
[1] 18
[1] 20
```

Example 2.13:

```
x <- c("A", "B", "C", "D")
for (i in seq(x)) {
   print(x[i])
  }
```

```
[1] "A"
[1] "B"
[1] "C"
[1] "D"
```

Alternatively,

```
for (i in 1:4) print(x[i])
[1] "A"
[1] "B"
[1] "C"
[1] "D"
```

2.4.2.3 Repeat Loop

A *repeat loop* is used to iterate over a block of code a multiple number of times. There is no condition check in a repeat loop to exit the loop. There should be explicit conditions within the body of the loop and then a *break* statement to terminate the loop.

Example 2.14:

```
r <- 1
repeat {
print(r)
r = r+1
if (r == 10){
break
}
}
[1] 1
[1] 2
[1] 3
[1] 4
[1] 5
[1] 6
[1] 7
[1] 8
[1] 9
```

2.4.2.4 While Loop

A *while loop* is used to loop until a specific condition is met.

Example 2.15:

```
w <- 1
while (w < 10) {
print(w)
w = w+1
}

[1] 1
[1] 2
[1] 3
[1] 4
[1] 5
[1] 6
[1] 7
[1] 8
[1] 9
```

2.4.3 Built-In Functions in R

There are a lot of functions in R. We will concentrate on the numerical functions, character function, and statistical functions that are commonly used in creating or recoding variables. The tables below show various categories of built-in functions and their descriptions.

2.4.3.1 Numerical Functions

Table 2.3 gives the numerical functions used in R and their descriptions with simple examples.

Table 2.3: Some numerical functions, descriptions, and examples.

Function	Description	Example
abs(x)	Absolute value	abs(-0.127) [1] 0.127
sqrt(x)	Square root	sqrt(200) [1] 14.14214
ceiling(x)	Numbers are rounded up to the nearest integer.	ceiling(100.125) [1] 101
floor(x)	Numbers are rounded down to the nearest integer.	floor(100.125) [1] 100
trunc(x)	Truncation	trunc(100.125) [1] 100
round(x, digits=n)	Rounding up	round(100.125, digits=1) [1] 100.1
signif(x, digits=n)	Significant figure	signif(100.125, digits=1) [1] 100
cos(x), sin(x), tan(x)	Trigonometric functions	cos(90) [1] -0.4480736
log(x)	Natural logarithm	log(90) [1] 4.49981
log10(x)	Common logarithm	log10(90) [1] 1.954243
exp(x)	Exponential	exp(0.125) [1] 1.133148

2.4.3.2 Character Functions

Character functions accept characters as an input and can return either character or numerical values as output. The character functions used in R and their descriptions with simple examples are shown Table 2.4.

Table 2.4: Some character functions, descriptions, and examples.

Function	Description	Example
substr(*x*, start=*n1*, stop=*n2*)	Extract or replace substrings in a character vector.	*x <- "peter"* *> substr(x, 2, 4)* *[1] "ete"*
grep(*pattern*, *x*, ignore.case=FALSE, fixed=FALSE)	Search for pattern in x. If fixed =FALSE then pattern is a regular expression. If fixed=TRUE then pattern is a text string. Returns matching indices.	*grep("Z", c("X","Y","Z"), fixed=TRUE)* *[1] 3*
sub(*pattern*, *replacement*, *x*, ignore.case =FALSE, fixed=FALSE)	Find pattern in x and replace it with replacement text. If fixed=FALSE then pattern is a regular expression. If fixed = TRUE then pattern is a text string.	*sub("\\s",".","Lord God")* *[1] "Lord.God"*
strsplit(*x*, *split*)	Split the elements of a character vector *x* at *split*.	*strsplit("faith", "")* *[[1]]* *[1] "f" "a" "i" "t" "h"*
paste(..., sep="")	Concatenate strings after using *sep* string to separate them.	*paste("child",1:3,sep=" ")* *[1] "child 1" "child 2" "child 3"*

2.4.3.3 Statistical Probability Functions

Table 2.5 shows the reserved statistical probability functions in R with descriptions and examples.

2.4.3.4 Other Statistical Functions

Table 2.6 shows the other statistical functions in R, the descriptions, and examples.

2.4.3.5 Other Useful Functions

Table 2.7 contains some useful functions, descriptions, and examples.

Table 2.5: Some statistical probability functions, descriptions, and examples.

Function	Description	Example
dnorm(*x*)	Normal density function (by default m=0 sd=1).	*dnorm(0.5)* *[1] 0.3520653*
pnorm(*q*)	Cumulative normal probability for q (area under the normal curve to the left of q).	*pnorm(1.96)* *[1] 0.9750021*
qnorm(*p*)	Normal quantile; value at the p percentile of a normal distribution	*qnorm(.75) # 75th percentile* *[1] 0.6744898*
rnorm(*n*, m=0,sd=1)	n random normal deviates with mean m and standard deviation sd.	*#50 random normal variates with* *mean=2.5, sd=0.05* *x <- rnorm(50, m=2.5, sd=0.05)*
dbinom(*x, size, prob*) **pbinom(*q, size, prob*)** **qbinom(*p, size, prob*)** **rbinom(*n, size, prob*)**	Binomial distribution where size is the sample size and prob is the probability of a heads (pi).	*# prob of 0 to 5 heads of fair coin out* *of 10 flips* *dbinom(0:5,10,0.5)* *# prob of 5 or less heads of fair coin* *out of 10 flips* *pbinom(5, 10, .5)*
dpois(*x, lambda*) **ppois(*q, lambda*)** **qpois(*p, lambda*)** **rpois(*n, lambda*)**	Poisson distribution with m=std=lambda.	*#probability of 0,1, or 2 events with* *lambda=4* *dpois(0:2, 4)* *[1] 0.01831564 0.07326256* *0.14652511* *# probability of at least 3 events with* *lambda=4* *1-ppois(2,4)* *[1] 0.7618967*
dunif(*x*, min=0, max=1) **punif(*q*, min=0, max=1)** **qunif(*p*, min=0, max=1)** **runif(*n*, min=0, max=1)**	Uniform distribution, follows the same pattern as the normal distribution above.	*#5 uniform random variates* *y <- runif(5)* *y* *[1] 0.14192169 0.13701585* *0.06418781 0.58657717 0.20230663*

2.4.4 User-Written Functions

R is flexible enough to allow the user to add functions. Some functions in R are functions of functions.

Table 2.6: Some statistical functions, descriptions, and examples.

Function	Description	Example
mean(x, trim=0, na. rm=FALSE)	Mean of object x	*# trimmed mean, removing any # missing values and 5 percent #of highest and lowest scores* *mx <- mean(x,trim=.05,na.rm=TRUE)*
sd(x) **var(x)** **mad(x)**	Standard deviation of object (x), var(x) for variance, and mad(x) for median absolute deviation.	*x<-1:10* *sd(x)* *[1] 3.02765* *var(x)* *[1] 9.166667* *mad(x)* *[1] 3.7065*
median(x)	Median	*x<-1:10* *median(x)* *[1] 5.5*
quantile(x, probs)	Quantiles where x is the numeric vector whose quantiles are desired and probs is a numeric vector with probabilities in [0,1].	*# 10th and 75th percentiles of x* *y <- quantile(x, c(0.1,0.75))* *y* *10% 75%* *1.90 7.75*
range(x)	Range	*range(x)* *[1] 1 10*
sum(x)	Sum	*sum(x)* *[1] 55*
diff(x, lag=1)	Lagged differences, with lag indicating which lag to use.	*diff(x, lag=1)* *[1] 1 1 1 1 1 1 1 1 1*
min(x)	Minimum	*min(x)* *[1] 1*
max(x)	Maximum	*max(x)* *[1] 10*
scale(x, center=TRUE, scale=TRUE)	Center Column or standardize a matrix	*x <- matrix(1:10, ncol = 2)* *centered_x* *[,1] [,2]* *[1,] -1.2649111 -1.2649111* *[2,] -0.6324555 -0.6324555* *[3,] 0.0000000 0.0000000* *[4,] 0.6324555 0.6324555* *[5,] 1.2649111 1.2649111* *attr(,"scaled:center")* *[1] 3 8* *attr(,"scaled:scale")* *[1] 1.581139 1.581139*

Table 2.7: Other useful functions, descriptions, and examples.

Function	Description	Example
seq(*from, to, by*)	Generate a sequence.	*gen <- seq(1,10,2)* *gen* *[1] 1 3 5 7 9*
rep(*x, ntimes*)	Repeat *x* *n* times.	*r <- rep(1:3, 2)* *r* *[1] 1 2 3 1 2 3*
cut(*x, n*)	Divide continuous variable in factor with *n* levels.	*y <- cut(x, 5)*

The structure of a function is:

my.function <- function(arg1, arg2, ...) {

statements

return(object)

}

Example 2.16:

```
my.result <- function(x,npar=TRUE,print=TRUE)
{ if (!npar) {
 center <- mean(x); spread <- sd(x)
} else {
 center <- median(x); spread <- mad(x)
}
if (print & !npar) {
  cat("Mean=", center, "\n", "SD=", spread, "\n")
} else if (print & npar) {
  cat("Median=", center, "\n", "MAD=", spread, "\n")
}
result <- list(center=center,spread=spread)
return(result)
}

# actualizing the function
set.seed(100)
rv1 <- rbinom(10, 3, 0.5)
rv2 <- my.result(rv1)
Median= 1
 MAD= 0
```

2.4.5 Importing, Reporting, and Writing Data

In this subsection, we discuss how to import packages and change directories for reading and writing local flat files, reading and writing local Excel files, connection interfaces, reading XML/HTML files, reading and writing to JSON, connecting to a database, and the textual format.

2.4.5.1 Packages

For anything you can think of doing with R, there is a package that has probably been written to execute it. The list of packages can be found in the official repository CRAN: http://cran.fhcrc.org/web/packages/.

The installation of any package is very easy in R—you use the command: *install.packages("packagename")*.

2.4.5.2 Working Directory and R Script

To read or write files from a specific location, we need to set a working directory in R. This example shows how to set the working directory in R to the folder "MyWorkingDirectory" on the C drive. Set the working directory from the menu if using the R-GUI (Change dir ...) or from the R command line:

```
setwd("C:\\MyWorkingDirectory")
setwd("C:/MyWorkingDirectory")  # can use forward slash
setwd(choose.dir( ))    # open a file browser
getwd( )   # returns a string with the current working directory
```

In addition, in order to have access and see a list of the files in the current directory, you can use the R command line:

```
dir ( )  # returns a list of strings of file names
dir (pattern = ".R$ ")   # list of files ending in ".R "
dir ("C:\\Users")    # show files in directory C:\Users

Run a script
source("helloworld.R")
```

2.4.5.3 Reading and Writing Local Flat Files

Local flat files can be read by using *read.table, read.csv*, or *readLines*. The acronym CSV stands for "comma separated values."

To illustrate, we will be working with the exchange rate dataset from a website. To procure this dataset, please visit:

http://bit.ly/MustaphaAbiodunAkinkunmi-BusinessStatisticsWithSolutionsInR-exchrt_csv

Right click and select "save as" to save it to your local directory. The code samples in the remainder of this section assume that you have set your working directory to the location of the exchrt.csv file. The R scripts below show how to read the exchrt.csv file from the local folder named "Business Statistics with Solutions in R" located in the C drive.

```
exch.rt<-read.csv("C: /Business Statistics with Solutions in R/exchrt.csv")
head(exch.rt, 7)
   Rate.Date    Currency Rate.  Year Rate.Month Buying.Rate Central.Rate Selling.Rate
1 10/4/2018    US DOLLAR       2018   October    305.4000    305.9000    306.4000
2 10/4/2018    POUNDS STERLING 2018   October    396.6230    397.2723    392.9217
3 10/4/2018    EURO            2018   October    351.1795    351.7544    352.3294
4 10/4/2018    SWISS FRANC     2018   October    307.6458    308.1495    308.6532
5 10/4/2018    YEN             2018   October      2.6766      2.6810      2.6854
6 10/4/2018    CFA             2018   October      0.5178      0.5278      0.5378
7 10/4/2018    WAUA            2018   October    424.2396    424.9341    425.6287

col.head<-readLines("C:/ Business Statistics with Solutions in R/exchrt.csv", 1)
col.head
[1] "Rate Date,Currency,Rate Year,Rate Month,Buying Rate,Central Rate,Selling Rate"
```

To write local flat files for the purpose of having a copy, you can used the functions such as *write.table, write.csv, writeLines*. It is advisable for someone to take note of the parameters before reading or writing a file.

```
write.table(exch.rt, "new_exchrt.csv")

Conversely, read.table is a    common function for reading data. Some of the argu-
ments of read.table( ) are described below:

function (file, header = FALSE, sep = "", quote = "\"'", dec = ".", numerals = c
("allow.loss", "warn.loss", "no.loss"), row.names, col.names, as.is = !
stringsAsFactors, na.strings = "NA",
  colClasses = NA, nrows = -1, skip = 0, check.names = TRUE,    fill = !blank.
  lines.skip, strip.white = FALSE, blank.lines.skip = TRUE, comment.char = "#",
  allowEscapes = FALSE, flush = FALSE,
  stringsAsFactors = default.stringsAsFactors(), fileEncoding = "", encoding = "un-
  known", text, skipNul = FALSE)
```

- **file** is the name of a file, or a connection.
- **header** is a logical value indicating the variables of the first line.
- **sep** is a string indicating how the columns are separated.
- **colClasses** is a character vector indicating the class of each column in the dataset.
- **nrows** are the number of rows in the dataset.
- **comment.char** is a character string indicating the comment character.
- **skip** is the number of lines to skip from the beginning.
- **stringsAsFactors** determines whether character variables are coded as factors.

2.4.5.4 Reading and Writing Excel Files

The functions *read.xlsx*, or *read.xlsx2*, can be used to read Excel files while *write.xlsx*, or *write.xlsx2*, can be used write Excel files. This is relatively fast but unstable. First, make sure that you install the Excel spreadsheet from the library. We can use the following commands to install the xlsx library.

```
# install the xlsx library
install.packages("xlsx")
```

The above command prompts the image in Figure 2.2. Click OK.

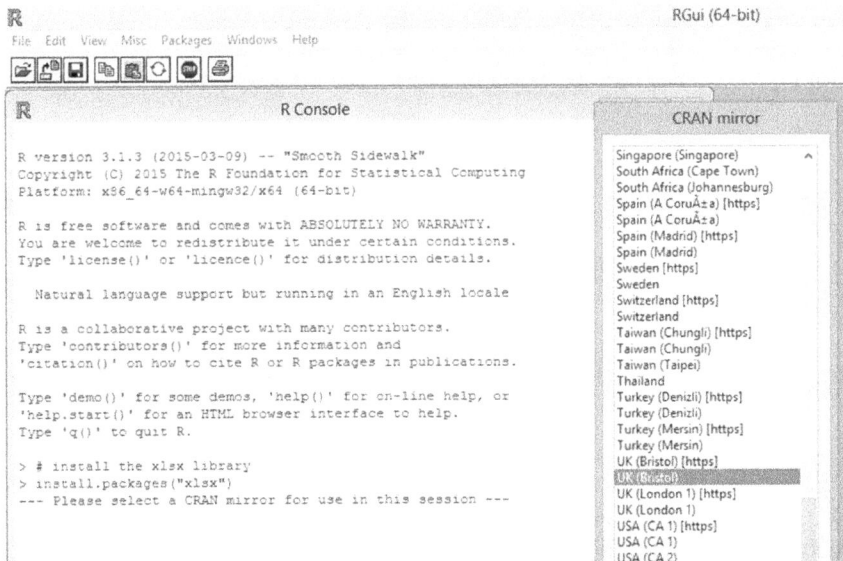

Figure 2.2: R software interface.

Second, we load the Excel spreadsheet into the library with the following command lines:

```
# load the library
library(xlsx)
```

Alternatively, you can load a library through the graphical interface:
- Click menu "Packages"
- Then select "Load package ... "
- Then select "xlsx"
- Click OK

2.4.5.5 Connection Interfaces

We may not necessarily need to access the connection interface directly; therefore connection may be made to files or to other channels such as:

 The function file is used to open a connection to a text file
- *The function gzfile* is used to open a connection to a file compressed with gzip
- *The function bzfile* is used to open a connection to a file compressed with bzip2
- *The function url* is used to open a connection to a webpage

The R code to read the data from the google drive is as follows:

```
library(data.table)
exch.rt<-fread("http://bit.ly/MustaphaAbiodunAkinkunmi-BusinessStatisticsWith
SolutionsInR-exchrt_csv")
head(exch.rt)
```

	Rate Date	Currency Rate	Year Rate	Month	Buying Rate	Central Rate	Selling Rate
1:	10/4/2018	US DOLLAR	2018	October	305.4000	305.9000	306.4000
2:	10/4/2018	POUNDS STERLING	2018	October	396.6230	397.2723	392.9217
3:	10/4/2018	EURO	2018	October	351.1795	351.7544	352.3294
4:	10/4/2018	SWISS FRANC	2018	October	307.6458	308.1495	308.6532
5:	10/4/2018	YEN	2018	October	2.6766	2.6810	2.6854
6:	10/4/2018	CFA	2018	October	0.5178	0.5278	0.5378

2.4.5.6 Connect to a Database

For good data management and an efficient warehouse, data should be stored in a relational database. A well-designed database architecture improves the robustness of how data can be mined to give a rigorous analysis. To perform statistical computing we will need very advanced and complex SQL queries. R can connect easily to many relational databases like MySQL, Oracle, and

SQL Server, among others. One of the advantages of using a database is that the records from the relational database are fetched as a data frame. Hence, it makes data crunching more easy to analyze with the aid of sophisticated packages and functions. R has some packages that connect relational databases to R, including: JSONPackages, RmySQL, RpostresSQL, RODBC, RMONGO, and others. In this book, we use MySQL as our reference database for connecting to R.

We can install the "RMySQL" package in the R environment using the following command:

```
install.packages("RMySQL")
library(RMySQL)
```

After installation, we will create a connection object in R to connect to the database. It requires the username, password, database name, and host name as input.

```
# Create a connection Object to MySQL database.
# assume our database named "my.database" that comes with MySql installation.
mysql.connect = dbConnect(MySQL(), user = 'root', password = '', dbname = ' my.database ',
  host = 'localhost')

# to list the tables available in this database
dbListTables(mysql.connect)
```

2.5 Data Exploration

Data exploration is the essential initial step to perform before the commencement of a proper analysis. This process involves data visualization (identifying summaries, structure, relationships, differences, and abnormalities in the data), descriptive statistics, and inferential statistics.

Often, no elaborate analysis is necessary since all the important conclusions required for a decision are evident from simple visual examination of the data. At other times, data exploration will be used to help guide the data cleaning, feature selection, and sampling process.

Look at the first three rows:

```
head(exch.rt[1:7], 3)
  Rate.Date  Currency Rate.Year Rate.  Month Buying.Rate Central.Rate  Selling.Rate
1 10/4/2018 US DOLLAR       2018 October    305.4000     305.9000      306.4000
2 10/4/2018 POUNDS STERLING 2018 October    396.6230     397.2723      392.9217
3 10/4/2018 EURO            2018 October    351.1795     351.7544      352.3294
```

For a more specific sample of the data, assuming that we want to view the last three columns of the data frame, and the first two rows in the columns. The command lines are:

```
# View the last three columns of the data frame, and the first two rows in those
columns
head(exch.rt[,5:7],2)
  Buying.Rate Central.Rate Selling.Rate
1    305.400    305.9000     306.4000
2    396.623    397.2723     392.9217
```

The *str()* function can be used to summarize the data frame. Let's look at the summary of the exchange rate data.

str(exch.rt)
'data.frame': 38728 obs. of 7 variables:
$ Rate.Date : Factor w/ 4126 levels "1/10/2002","1/10/2003",...: 623 623 623 623 623 623 623 623 623 623 ...
$ Currency : Factor w/ 27 levels "CFA","CFA ","DANISH KRONA",...: 21 12 5 19 25 1 23 27 13 18 ...
$ Rate.Year : int 2018 2018 2018 2018 2018 2018 2018 2018 2018 2018 ...
$ Rate.Month : Factor w/ 15 levels "8","April","August",...: 14 14 14 14 14 14 14 14 14 14 ...
$ Buying.Rate : num 305.4 396.62 351.18 307.65 2.68 ...
$ Central.Rate: num 305.9 397.27 351.75 308.15 2.68 ...
$ Selling.Rate: num 306.4 392.92 352.33 308.65 2.69 ...

2.5.1 Data Exploration through Visualization

In this section, we limit our scope to a few visual data exploration methods. These include a bar chart, a pie chart, and a box-plot (see Chapter 1).

2.5.1.1 Bar Chart

A bar chart (shown in Figure 2.3) is useful in visualizing a categorical data, and it is made up of rectangular bars with heights or lengths proportional to the values that they represent. The bars can be plotted vertically or horizontally.

Distribution of Age of Students

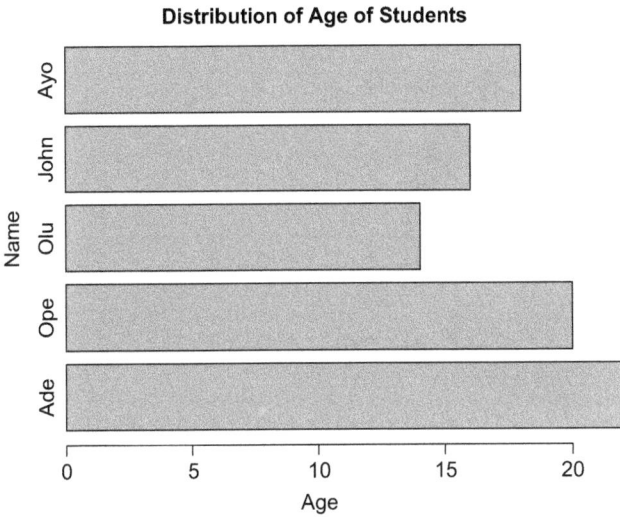

Figure 2.3: Bar chart for the distribution of students' ages.

```
# Simple bar chart
age <- c(22, 20, 14, 16, 18)
barplot(age, main = "Distribution of Age of Students", xlab = "Age", ylab
= "Name", names.arg = c("Ade", "Ope", "Olu", "John", "Ayo"), col = "green", horiz
= TRUE)
```

2.5.1.2 Pie Chart

A pie chart (shown in Figure 2.4) depicts the proportions and percentages be-tween categories by dividing a circle into proportional segments. Each arc length represents a proportion of each category. The full circle represents the total sum of all the data and is equal to 100%, represented in 360 degrees. A pie chart gives the reader an insight to the proportional distribution of a dataset. However, bar chart or dot plot are preferred to pie chart when you need to judge length more accurately.

Pie charts are created with the function *pie(x, labels=)* where x is a non-negative numeric vector indicating the area of each slice and labels= denotes a character vector of names for the slices.

```
# Age distribution of students
age <- c(22, 20, 14, 16, 18)
name <- c("Ade", "Ope", "Olu", "John", "Ayo")
```

Age Distribution

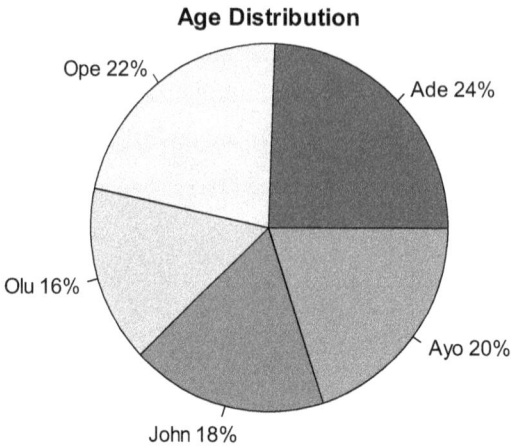

Ope 22%

Ade 24%

Olu 16%

Ayo 20%

John 18%

Figure 2.4: Pie chart for age distribution of students.

```
age.pct <- round(age/sum(age)*100)
name <- paste(name, age.pct) # add percents to labels
name <- paste(name,"%",sep="") # ad % to labels
pie(age, labels = name, col=rainbow(length(name)), main="Age Distribution")
```

2.5.1.3 Box-Plot Distributions

A box plot (shown in Figure 2.5) is very useful in showing the distribution or pattern between numerical and categorical attributes. Box plots can be created for individual variables or for variables by group. The format is *boxplot(x, data=)*, where x is a formula and data= denotes the data frame that is providing the data. A typical example of a formula is y~group where a separate box plot for a numerical variable y is generated for each value of a group. In this case, y is the response variable while group is the independent variable in the model. Add varwidth=TRUE to make box plot widths proportional to the square root of the samples sizes. Add horizontal=TRUE to reverse the axis orientation.

```
# simple box plot
age <- c(22, 20, 14, 16, 18)
gender= c("Female", "Male", "Male", "Male", "Female")
mydata<-data.frame(age, gender)
boxplot(age~gender,data=mydata, labels=c("Male","Female"), xlab="Gender",
ylab="Age", col=c("red","blue"))
```

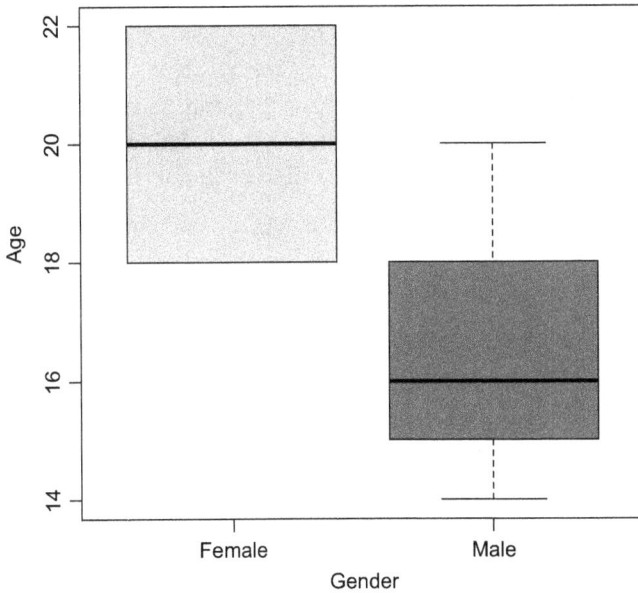

Figure 2.5: Box plot for gender.

Exercises for Chapter 2

1. What are the advantages of R software?
2. Explain the term vectorization in R using examples.
3. Enumerate the different types of data in R.
4. What do control structures in R allow you to do? List common control structures.
5. What the methods are used in visualizing data?

3 Descriptive Statistics

This chapter presents different statistical measures that can be employed to provide descriptive analysis of business-related data. We will define the statistical measures of central tendency and dispersion and use example data to show how they can be measured in R.

3.1 Central Tendency

Central tendency concerns the measures that may be used to represent the middle value of a set of data; that is, a value that represents the center of that data. The three common measures of central tendency are mean, median, and mode.

3.1.1 The Mean

The *mean* or *arithmetic mean* is the average of a set of observations. It is the aggregate of all observations divided by the number of observations in the data set. For instance, observations are denoted by $x_1, x_2, x_3, \ldots . x_n$. The sample mean is represented by $\bar{x}$ which is expressed as:

Mean of a sample:

$$\bar{x} = \frac{\sum_{i=1}^{n} x_i}{n} = \frac{x_1 + x_2 + x_3 + \ldots + x_n}{n} \tag{3.1}$$

Where $\sum$ is the summation symbol. The summation covers all of the data points.

For example, if you have 9 houses on your block and the number of people living in the houses is 2, 1, 6, 4, 1, 1, 5,1, and 3 and you want to know the average (mean) number of people living in a house, you simply add the numbers together and get 24 and divide that by 9 to get 2.67 people per house. In essence, we can say that any household in that block is expected to be made up of 3 people.

If the observation set covers a whole population, the symbol μ (the Greek letter *mu*) is used to represent the mean of the entire population. In addition, N is used instead of n to denote the number of elements. The mean of the population is specified as:

Mean of a population:

$$\mu = \frac{\sum_{i=1}^{N} x_i}{N} \tag{3.2}$$

https://doi.org/10.1515/9781547401475-003

Mean is the most commonly used measure of central tendency. In addition, the mean relies on information contained in all observations in a dataset.

Characteristics of mean:
- It summarizes all information in the dataset.
- It captures the average of all the observations.
- It is a single point viewed as the point where all observations are concentrated.
- All observations in the dataset would be equal to the mean if each observation has the same size or figure.
- It is sensitive to extreme observations (outliers).

3.1.2 The Median

The *median* is a special point because it is located in the center of the data, implying that half the data lies below it and half above it and is defined as a measure of the location of centrality of the observations. This is just an observation lying in the middle of the set.

In order to find the median by hand, given a small data set like this one, you need to reorder the dataset in numerical order:

1, 1, 1, 1, 2, 3, 4, 5, 6

If you divide the number of data points by two and round up to the next integer, you will find the median. So using the same example that we used to find the mean:

9/2 = 4.5

Rounded up the answer is 5, so the 5th data point is the median: 2.

This works nicely if you have an odd number of data points, but if you have an even number, then there would be two middle numbers. To resolve this dilemma, you take those two data points and average them. So let's add a house with 4 people. That means that the middle two numbers would be 2 and 3. So averaging them gives you 2.5 as the median.

Suppose that the monthly salaries of 10 workers in a manufacturing company are $3500, $3200, $2000, $2500, $2800, $3500, $1900, $2200, $3200, and $2800. Thus, the salary of a middle-class earner in the company is $2800.

Characteristics of median:
- It is an observation in the center of the dataset.
- One-half of the data lie above the observation.
- One-half of the data lie below the observation.
- It is resistant to extreme observations.

3.1.3 The Mode

The mode denotes the value that has the highest occurrence. Put differently, the mode is the value with most frequency in the dataset. In our example above, the mode would be 1 with four occurrences. Assume a baker sells wheat bread, flour bread, multi-grain bread and brown bread and the daily demand for the breads are 50 wheat breads, 15 flour breads, 28 multi-grain breads and 12 brown breads. This implies that people like to eat wheat breads most and she has an opportunity to increase the production of the wheat breads. Therefore, the mode of the sales of the breads is wheat breads and the probability that a customer will buy wheat bread is 0.48.

Example 3.1:
Calculate the mean, median, and mode of the following observations: 2, 6, 8, 7, 5, 3, 2, and 2.

Solution

To find the mean of the observations:
$$\bar{x} = \frac{2 + 6 + 8 + 7 + 5 + 3 + 2 + 2}{7} = \frac{35}{8} = 4.375$$
To solve for the mean of the observations in R, the following commands are expressed

```
x <- c (2,6,8,7,5,3,2,2)
 mean(x)
[1] 4.375
```

Finding the median of the observations requires the re-arrangement of the numbers from smallest to largest: 2,2,2,3,5,6,7 and 8. There are eight observations, so the median is the value in the middle, that is, in the fourth and fifth position. Those values are 3 and 5 so; you add the two numbers together and divide the outcome by 2 in order to get the median:

Median is $\frac{3 + 5}{2} = 4$ because 3 and 5 are the center of the dataset.

In R, the command is:

```
 median (x)
 [1] 4
```

See Figure 3.1 for a screen shot of this result.

The mode in the observation set is 2 because it appears three times while other values occur only once each.

There is no in-built function in R to calculate mode, but we can create a user function that takes the vector as input and gives the mode value as output.

In R, the user defined function to calculate mode looks like this:

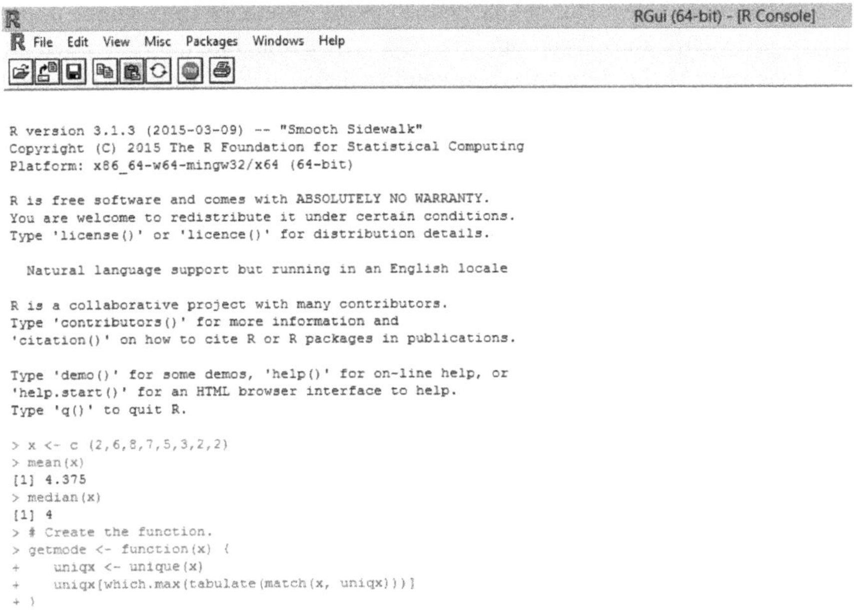
RGui (64-bit) - [R Console]
File Edit View Misc Packages Windows Help
R version 3.1.3 (2015-03-09) -- "Smooth Sidewalk"
Copyright (C) 2015 The R Foundation for Statistical Computing
Platform: x86_64-w64-mingw32/x64 (64-bit)
R is free software and comes with ABSOLUTELY NO WARRANTY.
You are welcome to redistribute it under certain conditions.
Type 'license()' or 'licence()' for distribution details.
Natural language support but running in an English locale
R is a collaborative project with many contributors.
Type 'contributors()' for more information and
'citation()' on how to cite R or R packages in publications.
Type 'demo()' for some demos, 'help()' for on-line help, or
'help.start()' for an HTML browser interface to help.
Type 'q()' to quit R.
> x <- c (2,6,8,7,5,3,2,2)
> mean(x)
[1] 4.375
> median(x)
[1] 4
> # Create the function.
> getmode <- function(x) {
+ uniqx <- unique(x)
+ uniqx[which.max(tabulate(match(x, uniqx)))]
+ }
Figure 3.1: Screen shot of calculations.

```
# Create the function.
getmode <- function(x) {
  uniqx <- unique(x)
  uniqx[which.max(tabulate(match(x, uniqx)))]
}
# Calculate the mode using the user function.
result <- getmode(x)
print(result)
```

3.2 Measure of Dispersion

There are different measures of variability or *dispersion*. These measure the degree that data is spread out from the mean of a dataset. These measures include range, interquartile range, variance, and standard deviation.

1. *Range*: This is the difference between the largest observation and the smallest observation. The range in Example 3.1 is the largest number – the smallest number = 8 – 2 = 6.

2. *Interquartile range*: This can be defined as the difference between the upper quartile (Q₃) and lower quartile (Q₁). Thus, the interquartile range is a measure of the difference between the higher half of your data and the lower half. To do this, we calculate the medians of the higher half and lower half and subtract the lower-half median from the higher-half median to get the interquartile range.

   – Using data from our first example, 1, 1, 1, 1, 2, 3, 4, 5, 6, we first need to split the dataset in two. So, we look to the middle and see that the number is odd, so we throw out the median (2). Then we find the median of 3, 4, 5 and 6 which is 4.5 and subtract the median of 1, 1, 1, 1 which is 1 to give us 3.5 as the interquartile range. So we see that the spread between the lower half of this data and the higher half is quite significant.

3. *Variance*: This is the average squared deviation of the data points from their mean.

   – The variance is found by calculating the difference from the mean for each data point, then squaring each of them, adding them all together, and then dividing by the number of items in the dataset minus one:

$$\frac{\sum_{i=1}^{n} (x_i - \bar{x})^2}{n - 1}$$

4. *Standard deviation*: This can be defined as the square root of the variance of the entire dataset. In financial analysis, the standard deviation is explored to capture the volatility and the risk associated with financial variables.

   – The standard deviation of a sample is the square root of the sample variance while the standard deviation of a population is the square root of the variance of the population. The formulas for these two forms of standard deviation are expressed below:

Sample standard deviation, $s = \sqrt{s^2} = \sqrt{\dfrac{\sum_{i=1}^{n} (x_i - \bar{x})^2}{n - 1}}$

Population standard deviation: $\sigma = \sqrt{\sigma^2} = \sqrt{\dfrac{\sum_{i=1}^{n} (x_i - \mu)^2}{n - 1}}$

Statisticians prefer working with the variance because its mathematical features make computations easy, whereas applied statisticians (those focusing on folding together domain and statistics through tools such as Six Sigma) like to work with the standard deviation because of its easy interpretation.

Let's calculate the variance and standard deviation of the data in Table 3.1. To compute this, a table is used for simplicity (before exploring R software to do everything within a minute).

In the above equation, the variance of the sample equals the sum of the third column, 41.8750, divided by n-1: $s^2 = \frac{41.875}{7} = 5.9821$. The standard deviation is the square root of the variance: $s = \sqrt{5.9821} = 2.4458$ or putting it in two-decimal places, s = 2.45.

There is shortcut formula for the sample variance that is equivalent to the formula above for variance:

$$s^2 = \frac{\sum_{i=1}^{n} x_i^2 - \left(\sum_{i=1}^{n} x_i\right)^2 \big/ n}{n-1}$$

Table 3.1: Observations for X variables.

x	$x_i - \bar{x}$	$(x_i - \bar{x})^2$
2	−2.375	5.6406
2	−2.375	5.6406
2	−2.375	5.6406
3	−1.375	1.8906
5	0.625	0.3906
6	1.625	2.6406
7	2.625	6.8906
8	3.625	13.1406
Total	0	41.8750

Solving the standard deviation and variance in Table 3.1 with the aid of R, the commands are:

```
x <- c(2,2,2,3,5,6,7,8)
var(x)
[1] 5.982143
sd(x)
[1] 2.445842
```

3.3 Shapes of the Distribution—Symmetric and Asymmetric

A symmetric distribution, also known as a normal distribution, describes variables occurring with regular or predictable frequency and with all central tendencies (mean, median and mode) found at the same point. Most people are familiar with this in the form of a bull curve. An asymmetric distribution consists of variables at irregular frequencies and with the three central tendencies found at different points. This is known as skewness (a degree of distortion away from the symmetric distribution). A normal distribution has skewness of zero, and an asymmetric distribution may contain either positive or negative skewness.

The shape of the distribution shows what the data looks like when it is graphed. The shape of the distribution gives insight into the characteristics of data and it tells us the appropriate descriptive statistics to be used. The shape of the distribution can be classified as follows:

1. *Symmetric distribution*: A dataset or a population is symmetric if one side of the distribution of the observation reflects a mirror image of the other; and if the distribution of the observations has only one mode, then the mode, the median, and the mean are the same.

2. *Asymmetric distribution*: This is the type of distributional pattern in which one side of the distribution is not a mirror image of the other. In addition, in its data distribution, the mean, median, and mode will not all be the same.

Additional attributes of a frequency distribution of a dataset are *skewness* and *kurtosis*.

3.3.1 Skewness Illustrated

Skewness measures the degree of asymmetry of a distribution. A right (positive) skewness occurs when the distribution stretches to the right more than it does to the left, while a left (negative) skewed distribution is one that stretches asymmetrically to the left. Graphs depicting a symmetric distribution, a right-skewed distribution, a left-skewed distribution, and a symmetrical distribution with two modes, are presented in Figure 3.2.

For a right-skewed distribution, the mean is to the right of the median which is to the right of the mode. The opposite is observed for left-skewed distribution (Figure 3.2). The calculation of skewness is reported by a number that may be positive, negative, or zero. Zero skewness indicates a symmetric distribution. Skewness could be different in terms of their shape even if two distributions have the same mean and variance.

3.3.2 Kurtosis

Kurtosis measures the peak level of the distribution. The larger the kurtosis, the *more* peaked the distribution. Its calculation is reported either as an absolute or

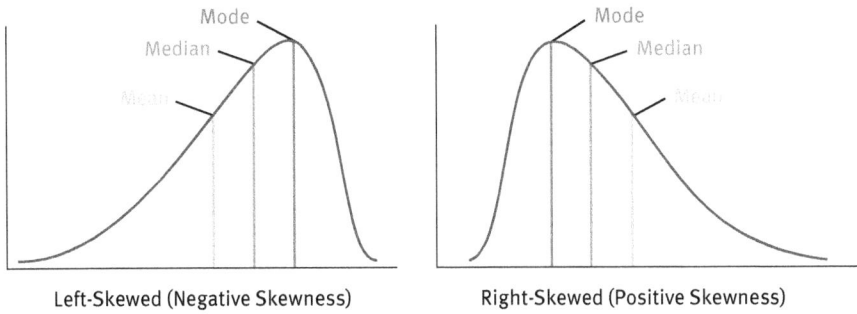

Left-Skewed (Negative Skewness)   Right-Skewed (Positive Skewness)

Figure 3.2(a): Skewness (left and right).

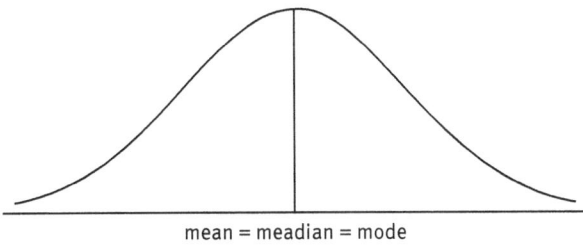

mean = meadian = mode

Figure 3.2(b): Symmetric distribution.

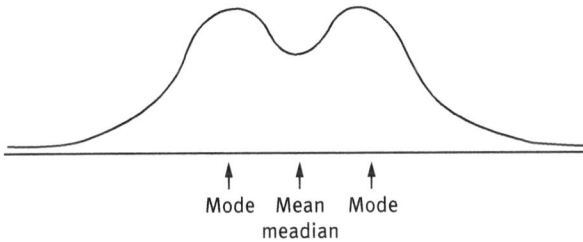

Mode   Mean   Mode
meadian

Figure 3.2(c): Asymmetric distribution.

a relative value. Absolute kurtosis is usually a positive number. For a normal distribution, the absolute kurtosis is 3. The value of 3 is used as the data point to compute relative kurtosis. Therefore, relative kurtosis is the difference between the absolute kurtosis and 3.

Relative kurtosis = absolute kurtosis − 3

The relative kurtosis can be either negative, known as *platykurtic* indicating a flatter distribution than the normal distribution; or positive, known as *leptokurtic* showing a more peaked distribution than the normal distribution.

Exercises for Chapter 3

1. Create a random variable X that includes all even numbers between 0 and 100. (e.g. 2, 4, ..., 100)
2. Calculate the mean, mode, and median of X.
3. Calculate the dispersion measures (standard deviation and variance) for X.
4. Explain the characteristics of central tendency measures (mean, mode, and median).
5. Explain kurtosis and skewness for shapes of distribution.

4 Basic Probability Concepts

In this chapter, we will discuss some basic concepts of probability, solving the problem of probability using Venn diagrams. The axioms and rules of probability will be discussed and will be extended to conditional probabilities. Practical examples with R code will be used for illustration.

4.1 Experiment, Outcome, and Sample Space

As we have mentioned before, it is important to understand the definitions of certain terms to be able to use them successfully. So, the first step in gaining an understanding of probability is to learn the terminology and the rest will be a lot simpler.

4.1.1 Experiment

Experiment is a measurement process that produces quantifiable results. Some typical examples of an experiment are: the tossing of a die, tossing of a coin, playing of cards, measuring weight of students, and recording growth of plants.

4.1.2 Outcome

Outcome is a single result from a measurement. Examples of outcomes are: getting a sum of 9 in the tossing of two dice, turning up of heads in the toss of a coin, selecting a spade from a deck of cards, and getting a weight above a certain threshold (say 50 kg).

4.1.3 Sample Space

The *sample space* is the set of all possible outcomes from an experiment and is denoted with S. The sample space of tossing a die is $S = \{1, 2, 3, 4, 5, 6\}$; the sample space of tossing a coin is denoted as $S = \{H, T\}$. In addition, the sample space for tossing two dice is

https://doi.org/10.1515/9781547401475-004

$$S = \begin{cases} \{1,1\}, \{1,2\}, \{1,3\}, \{1,4\}, \{1,5\}, \{1,6\}, \{2,1\}, \{2,2\}, \{2,3\}, \\ \{2,4\}, \{2,5\}, \{2,6\}, \{3,1\}, \{3,2\}, \{3,3\}, \{3,4\}, \{3,5\}, \{3,6\}, \\ \{4,1\}, \{4,2\}, \{4,3\}, \{4,4\}, \{4,5\}, \{4,6\}, \{5,1\}, \{5,2\}, \{5,3\}, \\ \{5,4\}, \{5,5\}, \{5,6\}, \{6,1\}, \{6,2\}, \{6,3\}, \{6,4\}, \{6,5\}, \{6,6\} \end{cases}$$

4.2 Elementary Events

Any subset of a sample set, empty set, and whole set inclusive is called an *event*. An *elementary event* is an event with a single element taken from a sample space and it is denoted as E. In addition, the elementary event can be written as $E \subset S$ (where E is a proper subset of the sample space, but not equal to the sample space). Suppose we have n possible outcomes from an experiment say E_1, E_2, E_3, ..., E_n, the space of elementary events from the sample space can be written as: $S = \{E_1 \cup E_2 \cup E_3 \cup \ldots \cup E_n\}$. This notation means that it is the sample is the union of all of the events. For instance, when tossing three coins, all possible outcomes are $\{HHH\}, \{HHT\}, \{HTH\}, \{THH\}, \{HTT\}, \{THT\}, \{TTH\}, \{TTT\}$, and when rolling a die, all possible outcomes are $\{1\}, \{2\}, \{3\}, \{4\}, \{5\}, \{6\}$. Hence, each of the outcomes are elementary events.

4.3 Complementary Events

Complementary events are events that cannot occur at the same time. Consider an event A, having its complement denoted by A' which are mutually exclusive and exhaustive. Hence, the sample space of the experiment can be written as $S = A \cup A'$ and also $\phi = A \cap A'$. That is, the intersection of A and its complement is the null set. For example, when tossing of a coin, turning a head and turning a tail are complementary events.

4.4 Mutually Exclusive Events

Two events are said to be *mutually exclusive* if they cannot occur at the same time. A good example of a mutually exclusive event is that when flipping a coin, either a head can appear or a tail, but the two cannot appear simultaneously. Mutually exclusive events A and B are denoted $A \cap B = \phi$.

4.5 Mutually Inclusive Events

Two events said to be *mutually inclusive*, if there are some common outcomes in the two events. Getting a head and a tail when flipping two coins is an example of mutually inclusive events. Another example is getting an odd number or prime number when throwing two dice.

4.6 Venn Diagrams

The use of Venn diagrams helps to visualize relations between sets. Figures 4.1–4.4 are diagrammatical representations of sets in a rectangular box.

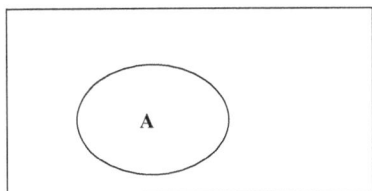

Figure 4.1: Venn diagram, A ⊂ S.

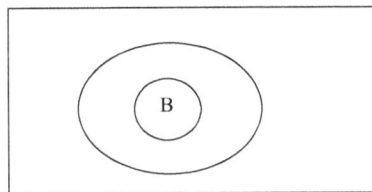

Figure 4.2: Two sets A and B, where B ⊂ A.

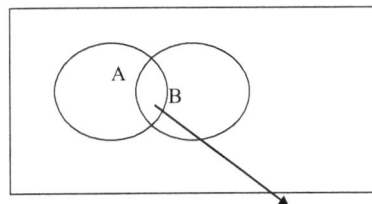

Figure 4.3: Two sets A and B, where A ∩ B.

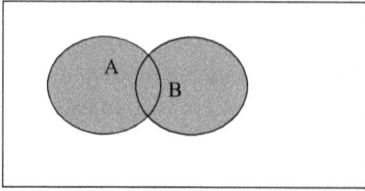

Figure 4.4: Two sets A and B, where A ∪ B is the shaded area.

4.7 Probability

The *probability* of event A can be defined as the number of ways event A may occur divided by the total number of possible outcomes. It is mathematically defined as:

$$\textbf{Probability (A)} = \frac{number\ of\ outcomes\ favorable\ to\ A}{number\ of\ possible\ \ outcomes}$$

or

$$P(A) = \frac{n(A)}{n(s)} \tag{4.1}$$

Example 4.1:
A fair die is rolled once, what is B = {1, 3, 5} the probability of: (1) rolling an even number? Or (2) rolling an odd number?

Solution

$$Probability(even\ number) = \frac{number\ of\ even\ number}{number\ of\ possible\ \ outcomes}$$

Let A be a set of even numbers, $A = \{2, 4, 6\}$, and B be a set of odd numbers, $B = \{1, 3, 5\}$, in tossing a die, with the sample space $S = \{1, 2, 3, 4, 5, 6\}$.
(1) $P(A) = \frac{3}{6}$
(2) $P(B) = \frac{3}{6}$

Example 4.2:
What is the probability that an applicant's resume will be reviewed within a week of submitting an application if 5,000 graduates applied for a job and the recruitment firm can only review 1,000 resumes in a week?

Solution

$$\text{Probability (treating a resume)} = \frac{1,000}{5,000} = 0.2$$

4.7.1 Simulation of a Random Sample in R

The R command *sample* is used to simulate drawing a sample. The function *sample* contains three arguments:

- *x* is a vector containing the objects
- *size* is the number of samples to be drawn
- *replace* is set to TRUE or FALSE depending on whether you want to draw the sample with replacement or not (see note).

Note: Sampling with replacement (replace = TRUE) indicates replacing the object before performing another sample selection. On the other hand, sampling without replacement (replace = FALSE) means selection in succession.

Example 4.3:
Consider an urn containing one blue ball, one green ball, and one yellow ball. To simulate drawing two balls from the urn with replacement and without replacement, we have:

```
# Urn contains one blue, one green, and one yellow ball
Urn = c('blue', 'green', 'yellow')

# Sampling with replacement
sample(x = Urn, size =2, replace = TRUE)
[1] "yellow" "yellow"

# Sampling without replacement
 Urn = c('blue', 'green', 'yellow')
 sample(x = Urn, size =2, replace = FALSE)
[1] "yellow" "blue"
```

For the two cases, there is no unique answer as long as the condition where they are selected is satisfied.
In the sampling with replacement, our result can be any from the sample space of:
["yellow" "yellow"], ["yellow" "blue"], ["blue" "yellow"], ["blue" "blue"], ["blue" "green"], ["green" "blue"], ["green" "yellow"], ["yellow" "green"], ["green" "green"]
However, in the case of sampling without replacement, we can have any of these outcomes:

["yellow" "blue"], ["blue" "yellow"], ["blue" "green"], ["green" "blue"], ["green" "yellow"], ["yellow" "green"]

So by putting the yellow ball back in after it was selected the first time, it expands the solution possibilities significantly. In this case it meant picking the yellow ball twice.

Example 4.4:
Draw 4 random numbers between 1 and 20 with replacement and without replacement.

```
# Sampling with replacement
 sample(x = 1:20, size =4, replace = TRUE)
[1] 17 7 7 10
```

```
# Sampling without replacement
 sample(x = 1:20, size =4, replace = FALSE)
[1] 1 17 18 20
```

Note: There is no unique answer for either case in as much as the condition of generating the samples is satisfied. It is generated randomly.

Example 4.5:
Simulate rolling a fair die.

```
# Sample from rolling a fair die
set.seed(25)
 sample(x = 1:6, size =1, replace = TRUE)
[1] 3
```

Note: The replacement option (TRUE or FALSE) does not hinder the outcome here since the experiment is performed once (i.e., size = 1). A seed is set in the code in Example 4.5 to make the code reproducible and to ensure the same random numbers will be generated each time the script is executed.

Example 4.6:
Suppose a company produces 6 types of cakes (butter cake, chiffon cake, genoise cake, pound cake, red velvet cake and sponge cake) with equal chances of production. The production manager intends to produce 100 cakes at random one after the other. What is the possible production sequence?

Solution
Let B, C, G, P, R and S represent butter cake, chiffon cake, genoise cake, pound cake, red velvet cake and sponge cake respectively.

```
# Sample replication
set.seed (25)
cake<-c("B", "C", "G", "P", "R", "S")
X<-replicate(100, sample(cake, size =1, replace = TRUE) )
 [1] "G" "R" "B" "S" "B" "S" "P" "G" "B" "C" "C" "G" "S" "P" "R" "B" "P" "R"
[19] "G" "R" "B" "G" "B" "B" "C" "C" "B" "P" "P" "C" "R" "G" "C" "B" "R" "G"
[37] "S" "G" "B" "B" "R" "C" "P" "R" "P" "C" "S" "R" "B" "R" "B" "R" "R" "G"
[55] "P" "B" "B" "P" "C" "B" "B" "C" "C" "R" "P" "S" "G" "S" "P" "C" "S" "B"
[73] "B" "R" "C" "G" "B" "S" "C" "C" "P" "G" "P" "B" "R" "R" "C" "G" "R" "P"
[91] "G" "C" "P" "P" "B" "G" "P" "S" "C" "C"mytable <- table(X)
mytable
X
 B C G P R S
22 16 13 18 14 17
```

This implies that 22% of the production is butter cake, 16% is chiffon cake, 13% is genoise cake, 18% is pound cake, 14% is red velvet cake and 17% is sponge cake.

4.8 Axioms of Probability

The expressions below are the axioms of probability:

- $P(A) \geq 0$, for all A (nonnegative)
  The probability of A is always greater than or equal to zero (nonegative).
- $P(s) = 1$ (normalization)
  The probability of the sample space is 1, meaning that probabilities of a sample will be between 0 and 1.
- If event A and event B are disjoint (i.e., $(A \cap B) = \emptyset$), then
  $P(A \cup B) = P(A) + P(B)$ (additivity)
  If event A is disjoint from event B, then the probability of both occurring is the same as adding the probability that event A will occur and the probability that event B will occur.
- For an infinite number of points, and $A_1, A_2, \ldots$ are disjoints, meaning they are mutually exclusive and cannot occur at the same time, then
  $P\left(\cup_{i=1}^{\infty} A_i\right) = \sum_{i=1}^{\infty} P(A_i)$. That is, for all infinite number of disjoint events A_i, the probability of all the events occurring is the same as the sum of the probability of the individual events A_i. However, the probability of occurrence at the same time is 0.

4.9 Basic Properties of Probability

The following are the basic properties of probability and are derived from the axioms of probability:

1. $P(A) = 1 - P(A')$
2. $P(A) \le 1$
3. $P(\phi) = 0$
4. $P(A) \le P(B)$ if only and if $A \subseteq B$; note that they can be equal
5. $P(A) = 1 - P(A') P(A \cup B) = P(A) + P(B) - P(A \cap B)$
6. $P(A \cup B) \le P(A) + P(B)$ or, alternatively, $P\left(\cup_{i=1}^{n} A_i\right) = \sum_{i=1}^{n} P(A_i)$

Suppose the probability that a pharmaceutical company will experience an increase in sales on a particular marketing campaign is 0.75, and the probability that the sales will remain unchanged is 0.25 then the following properties of probability hold:

Let A represents an event that sales will increase

Let B represents an event that sales will remain the same

then the $P(A) = 0 < 0.75 < 1$ and $P(B) = 0 < 0.25 < 1$; this satisfies properties (1) and (2)

since $P(A \cap B) = 0$, then $P(A \cup B) = P(A) + P(B) = 0.75 + 0.25 = 1$; this satisfies property (5)

4.10 Independent Events and Dependent Events

Two events A and B said to be *independent* if the occurrence of event A does not affect the occurrence of event B. In other words, two events A and B are statistically independent if:

$$P(A \ and \ B) = P(A) \times P(B)$$

Independent events can also be of the form:

$$P(A|B) = P(A) \ (and) \ P(B|A) = P(B)$$

So the probability of A given B is the same as the probability of only A happening, since A and B are independent of one another. Similarly, the probability of B given A is the same as the probability of only B happening, because they are independent events.

Any finite subset of events $A_{i1}, A_{i2}, A_{i3}, \ldots, A_{in}$ are said to be independent, if:

$$P(A_{i1}, A_{i2}, A_{i3}, \ldots, A_{in}) = P(A_{i1})P(A_{i2})P(A_{i3}) \ldots P(A_{in})$$

A typical example of independent events is flipping a coin twice, the outcome of head (H) or tail (T) facing up in the first toss does not affect the outcome in the second toss. Suppose a bank manager discovered that the probability that a customer prefers to open a saving account is 0.80. If three customers are selected at random, the probability that they will prefer saving accounts are independent events, since their choices of account do not affect one another.

However, two events A and B said to be *dependent* events if occurrence of event A affects the occurrence of event B or vice versa. The probability that B will occur given that A has occurred is referred to as *conditional probability* of B given A, and it can be written as $P(B|A)$.

Example 4.7:
A company is organizing a project team from 3 departments (the administrative department, the marketing department and the accounting department) with a total of 30 employees. There are 8 employees are in the administrative department, the marketing department has 12 employees and the accounting department has 10 employees. If two employees are selected to be on the team, one after the other:
(i) What is the probability that the first employee selected is from the accounting department and the second employee selected from administrative department if the first employee is also in the list of employees before the second employee is selected?
(ii) What is the probability that the first employee selected is from administrative department and the second is from marketing department if the selection is made without replacement?

Solution
A represents number of employees in the administrative department
M represents number of employees in the marketing department
C represents number of employees in the account department
(i) $P(C \text{ and } A) = P(C) \cdot P(A) = \frac{10}{30} \times \frac{8}{30} = 0.088$
(ii) $P(A \text{ and } M) = P(A) \cdot P(M|A) = \frac{8}{30} \times \frac{12}{29} = 0.11$

Example 4.8:
Table 4.1 shows the outcome of a survey conducted to look at the rate that small businesses fail despite the government programs directed at their survival. Calculate the probability that a restaurant is highly prone to the chance of this occurrence.

Table 4.1: Distribution of firms on their proneness to failure despite government support programs.

Proneness	Apparel/fashion	Restaurant	Technical	Total
Highly prone	216	14	20	250
Prone	30	10	5	45
Not prone	10	4	6	20
Total	256	28	31	315

Solution

$$P(restaurant|High\ prone) = \frac{number\ of\ high\ prone\ restaurants}{total\ number\ of\ high\ prone\ firms}$$

$$P(restaurant|High\ prone) = \frac{14}{250} = 0.056$$

4.11 Multiplication Rule of Probability

The *multiplication rule of probability* states that "the probability of the occurrence of two events is the probability of the intersection of the two events and is same as the product (multiplication) of the probability of the occurrence of one event and the probability of the occurrence of the second event."

$$P(A\ and\ B) = P(A \cap B) = P(A) \times P(B) \tag{4.2}$$

If and only if event A and event B are independent

$$P(A\ and\ B) = P(A \cap B) = P(A) \times P(B|A)$$

$P(B|A)$ indicates the probability of occurrence of event B given event A has occurred. Event A and event B are independent.

In general,

$$P(A_1 \cap A_2 \cap \ldots \cap A_n) = P(A_1)P(A_2|A_1)P(A_3|A_1 \cap A_2)\ldots P(A_n|A_1 \cap A_2 \ldots \cap A_{n-1}) \tag{4.3}$$

4.12 Conditional Probabilities

Let A and B be events in a sample, S, where the probability of A is not equal to zero (i.e., $P(A) \neq 0$), then the *conditional probability* that event B will occur given that event A has occurred is:

$$P(B|A) = \frac{P(A \cap B)}{P(A)} \qquad (4.4)$$

This is equivalent to:

$$P(A \cap B) = P(B|A) \times P(A) \qquad (4.5)$$

Conditional probability permits us to calculate probabilities of events based on partial knowledge of the outcome of a random experiment.

Example 4.9:
Consider that 10 non-defective products and 2 defective products are manufactured in one hour on a production line. To find the defective products, the store manager was randomly selecting the products one-by-one without replacement. What is the probability that the defective products are found in the first two selections?

Solution
Let D_1 be the event that the first selection is defective.
Let D_2 be the event that the second selection is defective.

$$P(D_1 \cap D_2) = P(D_1)P(D_2|D_1) = \frac{2}{10} \times \frac{1}{9} = 0.02$$

4.13 Computation of Probability in R

In this subsection, we will show how to use R commands and functions to compute probabilities. The following examples are given on how to calculate the corresponding probabilities for the outcomes, and then plot the graph of the relative frequency.

Example 4.10:
Consider rolling two fair dice. What is the probability of getting the sum of their outcomes when the experiment is performed 100 times? Plot the obtained results.

Step 1: Create a function to roll two fair dice and return their sum as follows:

```
dice.2 <- function () {
 dice <- sample (1:6, size = 2, replace =TRUE)
 return (sum(dice))
}
```

Step 2: Replicate the same experiment 100 times:

```
# replicate the experiment 100 times
set.seed (1010)
dice.roll<-replicate(100, dice.2( ))
 dice.roll
 [1] 6 7 11 10 8 11 10 2 8 3 8 4 3 4 9 5 8 5 6 4 7 9 5 9 3
[26] 11 4 3 8 2 6 12 9 9 6 10 4 4 6 3 8 9 6 8 5 6 9 8 6 11
[51] 8 4 2 8 7 7 8 10 5 12 9 11 11 5 8 7 6 3 5 6 7 7 5 12 8
[76] 6 9 7 8 12 12 4 12 5 10 7 3 4 6 2 7 9 6 9 3 6 2 7 11 6
```

Step 3: Tabulate the outcomes of the experiment:

```
# tabulate the outcomes
table (dice.roll)
dice.roll
2 3 4 5 6 7 8 9 10 11 12
5 8 9 9 15 11 14 11 5 7 6
```

Step 4: Calculate the probabilities:

```
# compute the probabilities of the sum of two dice
prob = table(dice.roll)/length(dice.roll)
prob
dice.roll
2 3 4 5 6 7 8 9 10 11 12
0.05 0.08 0.09 0.09 0.15 0.11 0.14 0.11 0.05 0.07 0.06
```

The probability values produced in R can be presented as:

Sum	2	3	4	5	6	7	8	9	10	11	12
Prob.	0.05	0.08	0.09	0.09	0.15	0.11	0.14	0.11	0.05	0.07	0.06

Step 5: Plot of relative frequency:

```
plot (table(dice.roll)/length(dice.roll), xlab = 'Sum of Outcomes ', ylab
= 'Relative Frequency', main = 'Graph of 100 Rolls of Two Fair Dice')
```

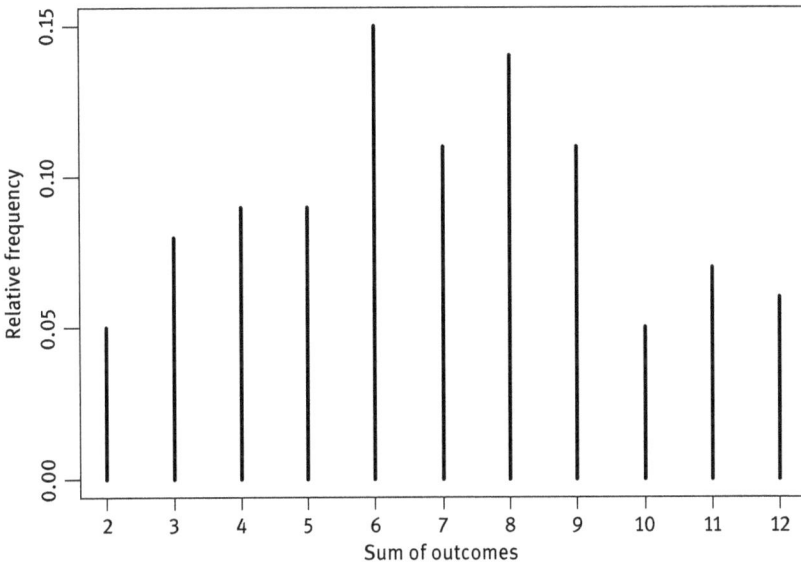

Figure 4.5: Graph of 100 rolls of two fair dice.

Example 4.11:

Consider a bottling company producing three (3) brands, namely, A, B, and C. The probability of producing A, B, and C is 0.54, 0.36, and 0.10. The probability of a defective product on the assembly line is 0.02 and non-defective product is 0.98. If a random sample of 1,000 units is taken of all products that are produced, what is the probability that the defective product is B?

Solution

```
# set seed number to always get the same result since the observation is randomly
generated
set.seed (1001)
Brands <- sample(c("A","B", "C"), 1000, prob=c(0.54, 0.36, 0.10), rep=TRUE)
Status <- sample(c("defective","non-defective"), 1000, prob=c(0.02,0.98),
rep=TRUE)
dataset <- data.frame(Status, Brands)
tabular <-with(dataset, table(Status, Brands))
tabular
```

```
          Brands
Status          A          B        C
  defective     11      7        0
  non-defective 536        357    89

# compute probability that B is defective
probB_D <- tabular[1, 2]/sum(tabular[1, ] )
probB_D
[1] 0.3888889
```

Exercises for Chapter 4

1. A fair die is tossed once. Calculate the probability that:
   a) exactly 3 comes up
   b) 3 or 5 come up
   c) a prime number comes up
2. Two dice are rolled together once. What is the probability that the sum of the outcome is:
   a) 4
   b) less than 4
   c) more than 4
   d) between 7 and 12 inclusive
3. A card is drawn from a well-shuffled pack of 52 cards. Find the probability of choosing:
   a) a king or a queen
   b) a black card
   c) a heart or a red king
   d) a spade or a jack
4. A research firm has 30 people on staff consisting of 2 research managers, 5 research associates, 3 administrative staff, and 20 fieldworkers. The Managing Director of the firm wants to set up a committee of 5 people. Find the probability that the committee will consist of:
   a) 1 research manager, 1 research associate and 3 fieldworkers
   b) 1 research manager, 2 research associates and 2 fieldworkers
   c) 2 research managers, 3 research associates
5. A firm has 100 employees (65 male and 35 female) and these employees were asked if they should adopt paternal leave or not. Their responses are summarized in a table below:

Gender	Favor	Oppose	Total
Male	62	3	65
Female	20	15	35
Total	82	18	100

a) Find the probability that a male is opposed to the paternal leave policy.
b) Given that a person is in favor of paternal leave policy, find the probability that the person is female.

5 Discrete Probability Distributions

A random variable (X) that takes integer values, $X_1, X_2, \ldots, X_n$ with the corresponding probabilities of $P(X_1)$, $P(X_1)$, $\ldots$, $P(X_n)$ and the probabilities $P(X)$ such that $\sum_1^n P(X) = 1$ is called a *discrete probability distribution*. The type of the random variable determines the nature of the probability distribution it follows. A discrete random variable is usually an integer value while a continuous random variable involves measuring and takes both integer values and a fractional part or real number. When the probabilities are assigned to random variables, then the collection of such probabilities give rise to a probability distribution. The probability distribution function can be abbreviated as *pdf*. In this book, we shall be using both the abbreviation and the full words. A discrete probability distribution satisfies two conditions:

$$0 \leq P(X) \leq 1 \text{ and} \tag{5.1}$$

$$\sum P(X) = 1 \tag{5.2}$$

5.1 Probability Mass Distribution

The probability that a discrete random variable X takes on a particular value x (i.e., $P(X = x) = f(x)$) is called a probability mass function (*pmf*) if it satisfies the following:

- $P(X = x) = f(x) > 0$, if $x \in S$

  All probability must be positive for every element x in the sample space S. Hence, if element x is not in the sample space S, then $f(x) = 0$.

- $\sum_{x \in S} f(x) = 1$

  The sum of probabilities for all of the possible x values in the sample space S must equal 1.

- $P(X \in x) = \sum_{x \in A} f(x)$

  The sum of probabilities of the x values in A is the probability of event A.

Example 5.1:
Experiment: Toss a fair coin 2 times
Sample space: $S = \{HH,\ HT,\ TH,\ TT\}$
Random variable X is the number of tosses showing heads
Thus $X{:}S \rightarrow \mathbb{R}$

https://doi.org/10.1515/9781547401475-005

$$X = (HH) = 2$$
$$X = (HT) = (TH) = 1$$
$$X = (TT) = 0$$

$X = \{0, \ 1, \ 2\}$. That is, random variable X takes a range of values 0, 1, and 2.
Hence, the probability mass function is given by:

$$P(X = 0) = \frac{1}{4}, P(X = 1) = \frac{1}{2}, \ \text{and} \ P(X = 2) = \frac{1}{4}$$

Example 5.2:
Suppose a real estate agent sold a number of houses in a month. The chances of selling no
house is 0.18, the chances of selling 1 house is 0.35, the chances of selling 2 houses is 0.42
and the chances of selling 3 houses is 0.05.

Let X be number of houses sold by the real estate agent.

The sample space for the sales of houses: X = {no house, 1 house, 2 houses and 3 houses}

Thus X takes the range of 0, 1, 2 and 3 and the corresponding probabilities are 0.18, 0.35,
0.42 and 0.05 respectively.

Therefore, the probability mass function can be written as:

$$P(X = x) = \begin{cases} 0.18 & \text{if } x = 0 \\ 0.35 & \text{if } x = 1 \\ 0.42 & \text{if } x = 2 \\ 0.05 & \text{if } x = 3 \\ 0, & \text{otherwise} \end{cases}$$

5.2 Expected Value and Variance of a Discrete Random Variable

The *expected value* is a measure of central tendency in a probability distribution
and the *variance* is a measure of dispersion of a probability distribution.

The mean or expected value (μ) for a discrete probability distribution function is computed as:

$$\mu = \sum\nolimits_{i=1}^{n} X_i \ P(X_i) \tag{5.3}$$

The variance (σ^2) of a discrete probability distribution function is computed as:

$$\sigma^2 = \sum\nolimits_{i=1}^{n} (X_i - \mu)^2 \ P(X_i) \tag{5.4}$$

The standard deviation (σ) of a discrete probability distribution function is given as:

$$\sigma = \sqrt{\sum\nolimits_{i=1}^{n} (X_i - \mu)^2 \, P(X_i)} \tag{5.5}$$

Examples of a discrete probability distribution are: rolling a die, flipping coins, counting car accidents on highways, producing defective and non-defective goods, number of sales made of a product, number of successful businesses established with a period of time, number of qualified applicants, etc.

The following are the common discrete probability distributions used in statistics: Bernoulli distribution, binomial distribution, geometric distribution, hypergeometric distribution, Poisson distribution, negative binomial distribution, and multinomial distribution.

Example 5.3:

A single toss of a fair die.

Roll (X)	1	2	3	4	5	6
P (X)	0.1667	0.1667	0.1667	0.1667	0.1667	0.1667

This has a discrete probability distribution since the random variable (X) takes the integers (i.e., 1, 2, 3, 4, 5, and 6) with corresponding probabilities of 0.1667 each. This satisfies that $0 \le P(X) \le 1$ and $\sum P(X) = 1$.

Example 5.4:

Consider a distribution of household size.

Assume an analyst obtained the result in Table 5.1 in a survey of 1000 households. Let the random variable X be the size of the number of households with probability of the outcome.

Table 5.1: Probability of household size.

Household size (X)	1	2	3	4	5	6+
P (X)	2/69	4/77	3/28	29/81	1/3	3/25

This satisfies that $0 \le P(X) \le 1$ and $\sum P(X) = 1$.

Example 5.5:

The number of defective items per month and the corresponding probabilities in a manufacturing firm are given in Table 5.2.

Table 5.2: Probability of defective items per month.

Defective(X)	0	1	2	3	4	5
P (X)	1/15	1/6	3/10	1/5	2/15	2/15

Since X is the number of defective items in a month, then equations $0 \le P(X) \le 1$ and $\sum P(X) = 1$ are satisfied.

Example 5.6:

Using the data in Example 5.5, calculate:
(i)   The expected value of the distribution
(ii) The standard deviation of distribution
(iii) The variance of the distribution

Solution

Accident (X)	0	1	2	3	4	5
p (X)	1/15	1/6	3/10	1/5	2/15	2/15
X p(X)	0	1/6	3/5	3/5	8/15	2/3
$X - \mu$	-2.57	-1.57	-0.57	0.43	1.43	2.43
$(X - \mu)^2$	6.59	2.45	0.32	0.19	2.05	5.92
p (X)(X - \mu)^2	0.44	0.41	0.10	0.04	0.27	0.79

(i) $\mu = \sum_{i=1}^{5} X_i \, P(X_i)$

Expected value = 0 + 1/6 + 3/5 + 3/5 + 8/15 + 2/3 = 2.57

(ii) Standard deviation

$$\sigma = \sqrt{\sum_{i=1}^{6} (X_i - \mu)^2 \, P(X_i)}$$

$$\sigma = \sqrt{0.44 + 0.41 + 0.10 + 0.04 + 0.27 + 0.79} = 1.43$$

(iii) Variance $(\sigma^2) = \sum_{i=1}^{n} (X_i - \mu)^2 \, P(X_i)$

$$\sigma^2 = 1.43^2 = 2.05$$

The code below shows how we can use R to get the solution to Example 5.6.

Using R code

```
# to calculate the mean, standard deviation, and variance of the distribution in
Example 5.6
set.seed (100)
x <- c(0, 1, 2, 3, 4, 5)
prob <- c(0.067, 0.167, 0.30, 0.20, 0.133, 0.133)
weighted.mean (x, prob)
[1] 2.56
x_miu = x - weighted.mean (x, prob)

# calculate the variance
x_miu2 = x_miu * x_miu
d = prob* x_miu2
var = sum(d)
var
[1] 2.045904

# calculate the standard deviation
sd = sqrt(var)
[1] 1.430351
```

5.3 Binomial Probability Distribution

In this section, we look at the one of the most common discrete probability distributions, its derivations, formula, statistical properties, and real life applications using examples. A couple of useful formulas shall be presented in this section to provide a better understanding of the binomial distribution. The derivation of the mean and variance of a binomial distribution shall be discussed.

A *binomial distribution* is a discrete probability distribution and is derived from the Bernoulli trial, a trial where there are two possible outcomes—"success" and "failure." The Bernoulli trial can be presented mathematically as:

$$X = \begin{cases} 1 \text{ if the outcome is a success} \\ 0 \text{ if the outcome is a failure} \end{cases} \tag{5.6}$$

Let p represent the probability of success, then 1-p will be the probability of failure. The probability mass function of X can be defined as:

$$f(x) = p^x(1-p)^{1-x}, \ x = 0 \ or \ 1 \tag{5.7}$$

However, the binomial probability function is the extension of the Bernoulli distribution when the following properties are met:
- Bernoulli trials are performed n times.
- Two outcomes, success (p) and failure (q) are possible in each of the trial.
- The trials are independent.
- The probability of success (p) remains the same between the trials.

Let X be the number of successes in the n independent trials, then the probability mass function of X is given as:

$$f(x) = \binom{n}{x} p^x (1-p)^{n-x}, \ x = 0, \ 1, \ \ldots, \ n \qquad (5.8)$$

$\binom{n}{x} = \dfrac{n!}{(n-x)!x!}$ reads: n combination x; and $n!$ reads: n factorial (e.g.,
$5! = 5 \times 4 \times 3 \times 2 \times 1$)
For example, $\binom{5}{2} = \dfrac{5!}{(5-3)!2!} = 10$

where n is the number of trials of the experiment performed, x is the number of successes recorded, p is the probability of success.

Example 5.7:
A production shift produces an engine per hour, and chances are one of the produced units will be defective is 0.75. If 10 hours are completed, what is the probability of 3 defective parts? Assume that number of defectives is binomially distributed.

Solution
Given that $n=10$, $x=3$, and $p=0.75$, the probability mass function of the binomial is given by:

$$f(x) = \binom{n}{x} p^x (1-p)^{n-x}, x = 0, \ 1, \ \ldots, n$$

$$f(x=3) = \binom{10}{3} (0.75)^3 (1-0.75)^{10-3}$$

$$f(x=3) = \frac{10!}{(10-3)!3!} (0.75)^3 (0.25)^7 = 0.0031$$

5.4 Expected Value and Variance of a Binomial Distribution

In this subsection, we will derive the expected value or mean of binomial distribution and also show how the variance of the binomial distribution is being derived.

5.4.1 Expected Value (Mean) of a Binomial Distribution

From the binomial probability mass function, we have:

$$f(x) = \binom{n}{x} p^x (1-p)^{n-x}, \, x = 0, \, 1, \, \ldots, n$$

Since the expected value of a discrete distribution is given as:

$$E(x) = \sum_{x=0}^{n} x.f(x)$$

Substituting for $f(x)$ gives:

$$E(x) = \sum_{x=0}^{n} x. \binom{n}{x} p^x (1-p)^{n-x}$$

Let

$$q = 1 - p$$

$$E(x) = \sum_{x=0}^{n} x. \binom{n}{x} p^x q^{n-x}$$

Now, substituting for the binomial coefficient yields:

$$E(x) = \sum_{x=0}^{n} x. \frac{n!}{x!(n-x)!} p^x q^{n-x}$$

$$E(x) = \sum_{x=0}^{n} x. \frac{n(n-1)!}{x(x-1)!(n-x)!} p^{x-1} pq^{n-x}$$

So,

$$E(x) = \sum_{x=0}^{n} x. \frac{n(n-1)!}{x(x-1)!(n-x)!} p^{x-1} pq^{n-x}$$

Pulling out np yields:

$$= np \sum_{x=1}^{n} \frac{(n-1)!}{(x-1)!(n-x)!} p^{x-1} q^{n-x}$$

and

$$= np \sum_{x-1=0}^{n-1} \binom{n-1}{x-1} p^{x-1} q^{(n-1)-(x-1)}$$

Since the sum of probabilities for all of the possible x values in the sample space S is equal to 1, then the result yields:

$$E(x) = np \tag{5.9}$$

Therefore, the expected value (mean) of the binomial probability function is np where n is the number of trials in the experiment and p is the probability of success.

5.4.2 Variance of a Binomial Distribution

The variance of a distribution can be computed from the expected value of that distribution.

$$var\ (x) = E(x^2) - (E(x))^2$$

Let
$$x^2 = x(x-1) + x.$$

Then, substitute for the value of x^2:

$$var\ (x) = E(x(x-1) + x) - (E(x))^2$$

Let's solve $E(x(x-1))$ using the same approach we used to derive the mean $(E(x))$ of a binomial distribution:

$$E(x(x-1)) = \sum_{x=0}^{n} x(x-1) \cdot \frac{n!}{x!(n-x)!} p^x (1-p)^{n-x}$$

So,

$$E(x(x-1)) = \sum_{x=0}^{n} x(x-1) \cdot \frac{n(n-1)(n-2)!}{x(x-1)(x-2)!(n-x)!} p^x (1-p)^{n-x}$$

Let
$$p^x = p^{x-2+2} = p^2 p^{x-2}.$$

Substituting for p^x gives:

$$E(x(x-1)) = \sum_{x=0}^{n} x(x-1) \cdot \frac{n(n-1)(n-2)!}{x(x-1)(x-2)!(n-x)!} p^2 p^{x-2} (1-p)^{n-x}$$

And substituting for $q = 1-p$ gives:

$$E(x(x-1)) = \sum_{x=0}^{n} x(x-1) \cdot \frac{n(n-1)(n-2)!}{x(x-1)(x-2)!(n-x)!} p^2 p^{x-2} q^{n-x}$$

Pulling out the constant terms and then evaluating yields:

$$E(x(x-1)) = n(n-1)p^2 \sum_{x=0}^{n} \frac{(n-2)!}{(x-2)!(n-x)!} p^{x-2} q^{n-x}$$

Since
$$\sum_{x=0}^{n} \frac{(n-2)!}{(x-2)!(n-x)!} p^{x-2} q^{n-x} = 1$$

Then,
$$E(x(x-1)) = n(n-1)p^2$$

$$var\ (x) = E(x(x-1)) + E(x) - (E(x))^2$$

Therefore, we can find the variance by substituting for $E(x(x-1))$.

$$var\ (x) = n(n-1)p^2 + np - (np)^2$$

$$var\ (x) = n^2 p^2 - np^2 + np - n^2 p^2$$

$$var\ (x) = np - np^2$$

$$var\ (x) = np(1-p),\ \text{or}$$

$$var\ (x) = npq \tag{5.10}$$

where n is the number of trials, p is the probability of success, and q is the probability of failure.

Here, we have shown that the variance of a binomial distribution is the
product of the number of trials(n), the probability of success(p), and the probability of failure (q).

Example 5.8:
A manufacturing company has 100 employees and an employee has a 5% chance of being absent from work on a particular day. It is assumed that the absence of a particular worker
would not affect another. The company can continue production on a particular day if no more
than 20 workers are absent for that day. Calculate the probability that out of 10 workers randomly selected, 3 workers will be absent. Hence, find the expected number of workers that
will be absent from work on that particular day.

Solution
(i)   Number of trials: $n = 10$ workers
      Possible outcomes: success (p) is the probability that a worker is absent from work, and
      failure (q) is the probability that a worker is not absent from work.
      Probability of success: p (worker is absent from work) = 0.05 and it is constant through all
      trials.
      The events are independent, i.e., the presence of a worker does not affect another.
      The probability mass function of the binomial is:

$$f(x) = \binom{n}{x} p^x (1-p)^{n-x},\ x = 0,\ 1,\ \dots,\ n$$

$$f(x=3) = \binom{10}{3} (0.05)^3 (1-0.05)^{10-3}$$

$$f(x=3) = \binom{10}{3}(0.05)^3(0.95)^7 = 0.0105$$

(ii) The expected numbers of workers to be absent on that day is: $np = 10 \times 0.05 = 0.5$. This is interpreted such that only one worker is expected to absent on that particular day.

Example 5.9:
An airline operator has twelve (12) airplanes. On a rainy day, the probability that an airplane will fly is 0.65. What is the probability on a rainy day that:
(i) an airplane will fly
(ii) three (3) airplanes will fly
(iii) at most two airplanes will fly
(iv) at least two airplanes will fly
(v) calculate the number of airplanes that expected to fly

Solution
(i) Let x be the number of airplanes to fly
    $n = 10$ and $p = 0.65$

$$f(x=1) = \binom{10}{1}(0.65)^1(0.35)^9 = 0.0005$$

(ii) $$f(x=3) = \binom{10}{3}(0.65)^3(0.35)^7 = 0.0212$$

(iii) $$f(x \le 2) = (f(x=0) + f(x=1) + f(x=2))$$

$$f(x \le 2) = \binom{10}{0}(0.65)^0(0.35)^{10} + \binom{10}{1}(0.65)^1(0.35)^9 + \binom{10}{2}(0.65)^2(0.35)^8$$

$$= 0.000028 + 0.0005123017 + 0.004281378 = 0.0048$$

(iv) $$f(x \ge 2) = 1 - (f(x=0) + f(x=1) + f(x=2))$$

$$= 1 - 0.0048 = 0.9952$$

(v) The expected number airplanes to fly on a rainy day is $np = 10 \times 0.65 = 6.5$. This indicates that only seven (7) airplanes will fly on a rainy day.

Example 5.10:
A marketing representative makes a sale on a particular product with a probability of 0.25 - per day. If he has 30 products to sell on a particular day, find the probability that:
(i) No sales are made
(ii) Five (5) sales are made
(iii) More than four (4) sales are made

Solution

(i)   Let X be number of sales made on the product

$$p(x = making\ sales) = 0.25$$
$$n = 30$$

$$p(x = 0) = \binom{30}{0}(0.25)^0(0.75)^{30} = 0.00018$$

(ii)

$$p(x = 5) = \binom{30}{5}(0.25)^5(0.75)^{25} = 0.1047$$

(iii)

$$p(x > 4) = 1 - [p(x = 0) + p(x = 1) + p(x = 2) + p(x = 3) + p(x = 4)]$$

$$= 1 - \left\{ \binom{30}{0}(0.25)^0(0.75)^{30} + \binom{30}{1}(0.25)^1(0.75)^{29} + \binom{30}{2}(0.25)^2(0.75)^{28} \right.$$

$$\left. + \binom{30}{3}(0.25)^3(0.75)^{27} + \binom{30}{4}(0.25)^4(0.75)^{26} \right\}$$

$$= 1 - \{0.00018 + 0.01786 + 0.008631 + 0.026853 + 0.06042\} = 0.90213$$

5.5 Solve Problems Involving Binomial Distribution Using R

In this subsection, we demonstrate how we can use R to generate a random sample from a binomial distribution through simulation and how to calculate probability from a binomial distribution parameter. We will use most of the examples illustrated in this chapter to compare the results with R outputs.

5.5.1 Generating a Random Sample from a Binomial Distribution

Here we use an R function to simulate binomial random variables with given parameters. The bar chart generated from binomial random samples is shown in Figure 5.1.

```
The function rbinom() is used to generate n independent binomial random variables.
The general form is:
rbinom (n, size, prob)
n is the number of random samples
prob is the probability of success
size is the number of trials
```

```
# To simulate 100 binomial random number with parameters n = 10 and p = 0.75
set.seed (150)
```

```
rv=rbinom (100, 10, 0.75)
rv
  [1] 10 8 8 6 9 8 8 8 8 6 7 5 6 8 7 5 7 9 9 8 8 9 10 9 8
 [26] 7 8 7 8 7 5 6 7 9 6 7 7 9 10 8 7 7 6 8 6 8 8 10 4 8
 [51] 6 6 6 7 9 8 5 9 8 6 6 5 10 8 7 8 6 8 8 6 7 4 6 8 10
 [76] 6 7 7 9 4 9 8 8 7 9 10 6 10 8 9 8 9 5 8 6 8 7 6 7 9

# the mean of the random sample
mean(rv)
[1] 7.44

# the standard deviation and variance of the random sample
sd(rv)
[1] 1.465564

var(rv)
[1] 2.147879
# To plot the bar chart of the random samples generated
set.seed (150)
rv=rbinom (100, 10, 0.75)
barplot(table(rv))
barplot(table(rv), ylab="Frequency", xlab="Random Number", main="X~Binomial (10,
0.75)")
```

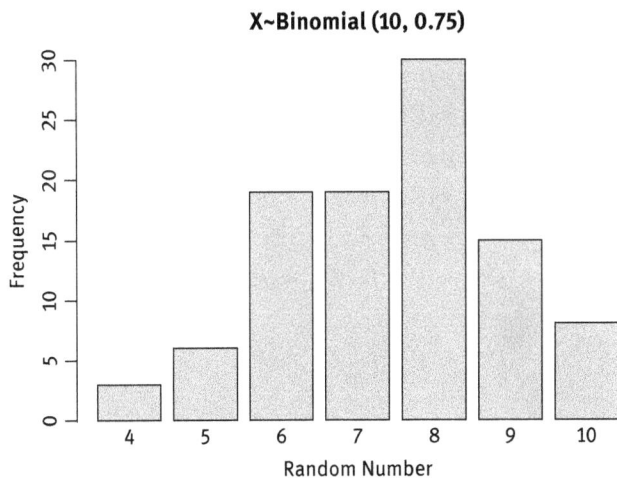

Figure 5.1: Bar chart of a Binomial random samples.

5.5.2 Calculate Probability from a Binomial Distribution

In this subsection, we are going to use R to also solve the binomial distribution problems in Examples 5.8 to Example 5.10 given above.

```
dbinom is the function to calculate a binomial probability distribution
function (x, size, prob, log = False)
x is the number of successes
size is the number of independent trials
prob is the probability of success
```

Example 5.11:
Using the data from Example 5.8, use R code to obtain the binomial probability distribution and compare the result.

```
dbinom(3,10,0.05)
[1] 0.01047506
This is exactly what we got in Example 5.8.
```

Example 5.12:
Use the data in Example 5.9 and use R to obtain the probabilities in Example 5.9 (i)–(iv) and compare the results.

```
# an airplane will fly
dbinom(1,10,0.65)
 [1] 0.0005123017

# three(3) airplanes will fly
dbinom(3,10,0.65)
[1] 0.02120302

# at most two airplanes will fly
no_air=dbinom(0,10,0.65)
air1=dbinom(1,10,0.65)
air2=dbinom(2,10,0.65)
prob_most2= no_air+ air1+ air2
prob_most2
 [1] 0.004821265

# at least two airplanes will fly
prob_least2=1-prob_most2
prob_least2
[1] 0.9951787
```

```
# Expected number airplanes to fly on a rainy day
p<-0.65
n<-10
expected_value<-n*p
> expected_value
[1] 6.5
```

Good! We got the same results.

Example 5.13:

Use the data in Example 5.10 to solve the example in R and then compare results.

```
# no sale of the product
 dbinom(0,30,0.25)
[1] 0.0001785821
```

```
# 5 sales of the products
 dbinom(5,30,0.25)
[1] 0.1047285
# more than 4 sales of the products
p0<-dbinom(0,30,0.25)
p0
[1] 0.0001785821
```

```
p1<-dbinom(1,30,0.25)
 p1
[1] 0.001785821
```

```
p2<-dbinom(2,30,0.25)
p2
[1] 0.008631468
```

```
p3<-dbinom(3,30,0.25)
p3
[1] 0.02685346
```

```
p4<-dbinom(4,30,0.25)
p4
[1] 0.06042027
```

```
prob<-1-(p0+p1+p2+p3+p4)
prob
[1] 0.9021304
```

The results in the Example 5.10 (i)-(iv) are the same as when we used R.

Exercises for Chapter 5

1. (a) What is a discrete probability distribution and what is it used for?
   (b) State the conditions that need to be satisfied for a discrete probability distribution.

2. (a) If X is a discrete random variable with the probability $p(X)$, what is the expected value and standard deviation of X?
   (b) The table below shows the chances of spoilage in a plant preparing chickens for sale to markets with their respective probabilities.

Spoilage Cases (X)	0	1	2	3	4	5	6	7	8
P (X)	0.12	0.20	0.15	0.18	0.12	0.08	0.07	0.03	0.05

   i.   Calculate the mean of the distribution of the spoilage cases.
   ii. Calculate the variance of the distribution.
   iii. Calculate the standard deviation of the distribution.

3. (a) List the names of the discrete probability distribution.
   (b) State the binomial probability distribution.
   (c) How do you derive the expected value and the variance of a binomial distribution?

4. The probability of contaminated products reported by customers is 0.45. If ten (10) customers purchase the particular products, what is the probability that:
   (a) One customer will purchase a contaminated product?
   (b) Two customers will purchase a contaminated product?
   (c) At most two customers will purchase a contaminated product?
   (d) At least two customers will purchase a contaminated product?
   (e) With this information, calculate the number of customers that can be expected to purchase contaminated products.

5. The Corporal Affairs Commission discovered that half of the unregistered firms within a geographical location were distressed per year. If 10 unregistered firms are selected in the region, what is the probability that:
   (a) None of the unregistered firms in the region will be in distress
   (b) Less than or equal to 1 unregistered firm will be in distress
   (c) More than 2 unregistered firms will be in distress
   (d) Less than or equal to 3 unregistered firms will be in distress

6 Continuous Probability Distributions

In the previous chapter, we discussed discrete probability distributions and their properties. In the discrete probability distribution, the random variable takes only integer values or a countably infinite number of possible outcomes. However, in a *continous probability distribution*, the random variable takes any real value within a specified range. Typical examples of continous random variables are weight, temperature, height, and some economic indicators (prices, costs, sales, inflation, investments, and so on). A continuous probability distribution demonstrates the complete range of values a continuous random variable can take with their associated probabilities. This distribution is very useful in predicting the likelihood of an event within a specified range of values. In this section, we will observe that the sum symbol, Σ, which is used in deriving the mean and variance of discrete probability distribution, has turned to an integral symbol, $\int$. The integral sign indicates the sum of continous random variables over an interval of points. Examples of continous probability distributions are the normal distribution, exponential distribution, student-t distribution, chi-square distribution, and so on.

If X is a continuous random variable that can take any real value within a specified range, then the probability over a random variable is a *continuous probability distribution*. Considering x to be continous, the probability density function denoted by $f(x)$ such that the probability of event $a \leq x \leq b$ is represented mathematically as:

$$P(a \leq x \leq b) = \int_{a}^{b} f(x)dx$$

where a and b are the limits or point intervals.

This implies the probability of a continous variable x within a specified range, say, a and b, is the same as taking the integral function of the random variable over its bounded range.

However, the probability is zero if the continous random variable x is not bounded by an interval.

That is,

$$P(x = a) = \int_{a}^{a} f(x)dx = 0$$

Since $b = a$

https://doi.org/10.1515/9781547401475-006

Example 6.1:
The time required by a driver to drive his boss from home to the office is a function of $\left(\frac{1}{x^2}\right)$. What is the probability that the driver will get to the office between 5 and 10 minutes?

Solution
x denotes the time required for the driver to travel from home to office, given that:

$$f(t) = \frac{1}{x^2}, a = 5 \text{ and } b = 10$$

Then,

$$P(5 \leq x \leq 10) = \int_5^{10} \left(\frac{1}{x^2}\right) dx$$

Integrate the right-hand side (RHS) over the interval,

$$P(5 \leq x \leq 10) = \left[-\frac{1}{x}\right]_5^{10}$$

Substitute the value of x and then evaluate the RHS,

$$P(5 \leq x \leq 10) = \left[-\frac{1}{10}\right] - \left[-\frac{1}{5}\right]$$

$$P(5 \leq x \leq 10) = 0.10$$

The probability that the driver will reach the office from home within 5 to 10 minutes is 0.10. This means that he has a low chance of getting to office under this condition.

6.1 Normal Distribution and Standardized Normal Distribution

In this subsection, we discuss how a *normal distribution* is the most important distribution because if you have many independent variables with non-normal distributions, the aggregation of these variables will tend to a normal distribution as the number of observations is large (i.e., central limit theorem, the observation that with a large enough sample drawn from a population with finite variance, the mean of all samples will equal the mean of the population). The central limit theorem is discussed in Chapter 7 of this book. A normal distribution's charateristics make it the most widely used distribution in statistics and applied mathematics. It is also useful in measuring moments, kurtosis, skewness, and so on. A normal distribution is also known as a *Gaussian distribution*, named after after the mathematician Karl Friedrich Gauss.

A normal distribution is a bell-shaped curve with the mean denoted by μ and standard deviation denoted by σ. The density curve of a normal distribution is symmetrical, centered about its mean, and spread by its standard deviation. The altitude of a normal density curve at a point y is given as:

$$f\left(y;\mu,\sigma^2\right) = \frac{1}{\sigma\sqrt{2\pi}}exp\left\{-\frac{1}{2}\left(\frac{y-\mu}{\sigma}\right)^2\right\}, \ -\infty \leq y \leq \infty$$

where y is a continuous random variable, μ is the mean, and σ^2 is the variance of the distribution.

The *standard normal curve* is a special case of the *normal curve* when the mean is zero and the standard deviation is 1. A normal distribution can be written in the form $X \sim N(\mu, \sigma^2)$ and reads: X is normally distributed with mean μ and variance σ^2. The total area under the curve is 100%, i.e., all observations fall under the curve. A dataset is said to be normally distributed, if 68.27% of the observations fall within $\pm 1SD$ of the mean. Also, about 95.45% of the observations will fall within $\pm 2SD$, and 99.73% of the observations will fall within $\pm 3SD$ as shown in Figure 6.1.

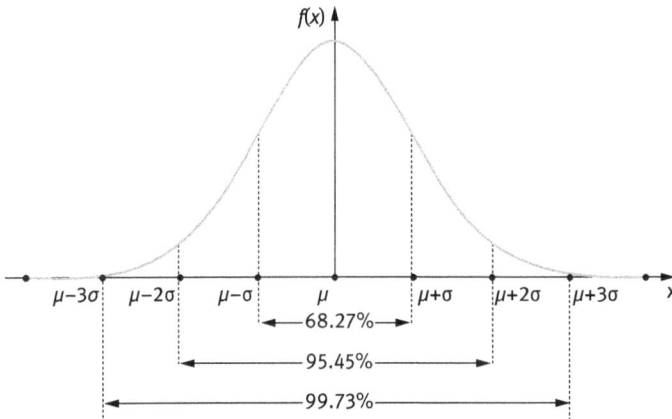

Figure 6.1: Normal curve.

6.1.1 Properties of a Normal Distribution

The following are the properties of a normal distribution:
1. The mean, median. and mode have the same value
2. The curve is symmetrical
3. The total area under the curve is 1
4. The curve is denser in the center and less dense in the tails
5. Normal distribution has two parameters, mean (μ) and variance (σ^2)

6. About 68% of the area of a normal distribution is within one standard deviation of the mean
7. About 95% of the area of a normal distribution is within two standard deviations of the mean
8. Over 99% of the area of a normal distribution is within three standard deviations of the mean

6.2 Standard Normal Score (Z-Score)

The *standard normal score* is the standardized value of a normally distributed random variable and it is usually referred to as a z-score. A standard normal score is useful in the area of statistics because it enables us to find the the probability of a score occurring within a normal distribution and also use it to make a comparison between two scores that come from different normal distributions. A random variable is standardized by subtracting the mean of the distribution from standardized value, and then dividing it by the standard deviation of the distribution. The z-score of a random variable can be written mathematically as:

$$z = \frac{X-\mu}{\sigma} \sim N(0,1)$$

This reads as: z is normally distributed with a mean of zero and standard deviation of one where X is the random variable to be standardized, μ is the mean of the distribution, and σ is the standard deviation of the distribution. Thus, once the random variable is standardized, it is approximately normal with a mean of 0 and a standard deviation of 1—which is the same as standard normal.

Example 6.2:
A class of 30 job applicants for an accounting position took an exam in mathematics with the class mean score of 65 and standard deviation of 12.5. Assuming that the scores are normally distributed, what percentage of applicants scored above 70 on the exam?

Solution
Given that: $n = 30$, $\mu = 65$ and $\sigma = 12.5$. $P(x > 70) = ?$
Standardize the applicants' score:

$$P(x > 70) = P\left(\frac{x-65}{12.5} > \frac{70-65}{12.5}\right)$$

Let
$$z = \frac{x-65}{12.5} = 0.4$$

$$P(x > 70) = P(z > 0.4)$$

Since sum of probability is 1, the left side of the normal curve can be written as:

$$P(x > 70) = 1 - P(z \leq 0.4)$$

Look at the probability corresponding to $z \leq 0.4$ from the standard normal table. That is, $\Phi(0.4) = 0.6554$

$$P(x > 70) = 1 - \Phi(0.4) = 1 - 0.6554 = 0.3446$$

Thus, 34% of the applicants that sat for the mathematics examination scored above 70. The sketch of the region is depicted in Figure 6.2.

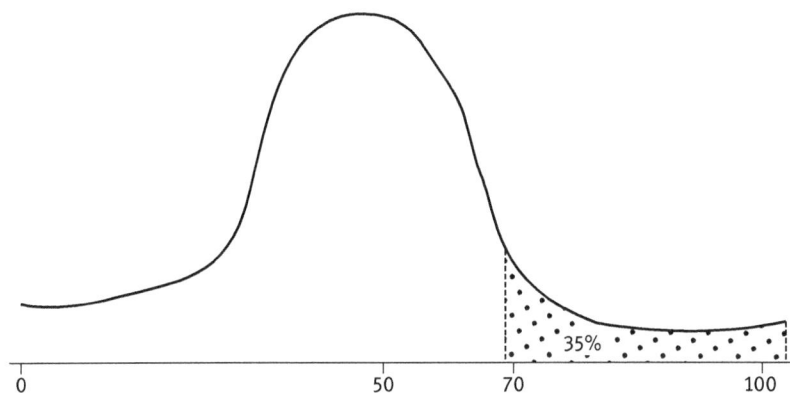

Figure 6.2: Sketch of Example 6.2.

Example 6.3:
A production machine produces plastics with a mean of 80 and standard deviation of 5 in one minute. Assuming that the production of plastics follows a normal distribution, calculate:
  (i) the probability that the machine will produce between 78 and 85 items?
  (ii) the probability that the machine will produce fewer than 90 items?

Solution
Given that: $\mu = 80$ and $\sigma = 5$
  First, standardize the production variable:
  Let

$$z = \frac{X - 80}{5}$$

(i)
$$P(78 \leq x \leq 85) = P\left(\frac{78 - 80}{5} \leq \frac{X - 80}{5} \leq \frac{85 - 80}{5}\right)$$

Substitute for
$$z = \frac{X - 80}{5}$$

$$P(78 \leq x \leq 85) = P\left(\frac{78-80}{5} \leq z \leq \frac{85-80}{5}\right) = P(-0.4 \leq z \leq 1)$$

Look for the probability corresponding to $-0.4 \leq z$ and $z \leq 1$ from the standard normal table and subtract the lowest value from highest value.

$$= \Phi(1) - \Phi(-0.4) = 0.8413 - (1 - 0.6554) = 0.4967$$

This means that the probability that the machine will produce between 78 to 85 plastics within a minute is 0.5.

(ii)   Standardize the production variable:

$$P(x < 90) = P\left(\frac{x-80}{5} < \frac{90-80}{5}\right) = \Phi(2) = 0.9772$$

$$P(x < 90) = P\left(z < \frac{90-80}{5}\right) = \Phi(2) = 0.9772$$

This indicates that the probability that the machine will produce fewer than 90 plastics is 0.98; this implies it is almost sure for the machine to produce up to 90 plastics within one minute.

Example 6.4:

Suppose that a Corporate Affairs Commission official knows that the monthly registration of new companies follows a normal distribution with an average of 50 new registrations per month and a variance of 16 new registrations. Find the probability that:

(i)   fewer than 35 new companies will register

(ii) 45 new companies will register

(iii) between 35 and 45 new companies will register

Solution

Let x be the number of registered companies,

x is normally distributed with mean (50) and variance (16). That is, $x \sim N(50, 16)$.

Since the standard deviation is the square root of the variance then,

(i) $P(x < 35) = P\left(\frac{x-50}{4} < \frac{35-50}{4}\right)$
      $P(x < 35) = P(z < -3.75) = 1 - \Phi(3.75) = 0.99991$

This means that it is very likely that fewer than 35 companies will register within a one month period.

(ii) $P(x = 45) = P\left(\frac{x-50}{4} = \frac{45-50}{4}\right) = 1 - \Phi(1.25) = 1 - (0.89435) = 0.1056$

This result shows that there is low probability of having exactly 45 companies register within one month.

(ii) $P(35 \leq x \leq 45) = P\left(\frac{35-50}{4} \leq z \leq \frac{45-50}{4}\right) = \Phi(-1.25) - \Phi(-3.75) = 0.10556$

This means there is low chance of having between 35 and 45 companies register within a one month period.

6.3 Approximate Normal Distribution to the Binomial

In this subsection, we learn how to use the normal distribution to approximate binomial probability. The central limit theorem is the tool that allows us to do this. Suppose that we have about 500 employees that are ready to take a promotion test in order to move to the next level. The manager of the personnel department estimated that 75% of the staff sitting for the test will pass the first time. If we want to find the probability that more than 300 employees will pass the promotion test the first time, then this is a binomial event with $n = 500, p = 0.75$ and $P(x > 350) =$.

It will be tedious calculating the required:

$$P(x = 350) + P(x = 351) + P(x = 352) + \ldots + P(x = 500)$$

$$\binom{500}{350}(0.75)^{350}(0.75)^{150} + \binom{500}{351}(0.75)^{351}(0.75)^{149}$$

$$+ \ldots + \binom{500}{500}(0.75)^{500}(0.75)^{0}$$

Due to the computational effort involved, we can instead use a normal approximation to solve this problem.

This general rule of thumb for the approximation is that the sample size (n) is "sufficiently large" (i.e., $np \geq 5$ and $n(1 - p) \geq 5$.

Since $np = 500 \times 0.75 = 375 \geq 5$, then we can apply a normal approximation.

Let $z = \frac{X - \mu}{\sigma}$

Recall that the mean of a binomal distribution is pq and the variance is npq.

We can redefine the z-score:

$$z = \frac{X - np}{\sqrt{npq}}$$

Thus,
$$\sigma^2 = npq = 500 \times 0.75 \times 0.25 = 93.75$$

and
$$\sigma = 9.68$$

Since $x > 350$ implies 350 not inclusive, it is possible to write an approximation value as $x \geq 350.5$

$$P(x > 350) \approx P(x \geq 350.5) = P\left(\frac{x - 375}{9.68} \geq \frac{350.5 - 375}{9.68}\right)$$

$$P(z \geq -2.53) = 1 - \Phi(-2.53) = 1 - 0.00570 = 0.9943$$

This is interpreted as it is almost certain that more than 350 staff taking the test will pass the first time.

6.4 Use of the Normal Distribution in Business Problem Solving Using R

To generate a normal distribution function, we use the function *pnorm*. The *dnorm, qnorm*, and *rnorm* functions are used for density, quantile function, and random generation for the normal distribution, respectively.

```
pnorm(q, mean =0, sd = 1, lower.tail = TRUE, log.p = FALSE)
dnorm(x, mean = 0, sd = 1, log = FALSE)
qnorm(p, mean = 0, sd = 1, lower.tail = TRUE, log.p = FALSE)
rnorm(n, mean = 0, sd = 1)
      where   q is the vector of quantiles
         mean is the vector of means
             sd is the vector of standard deviations
             log.p takes the logical value and if log.p = TRUE, then it means
                we are working with the logarithm of probabilities p, if
                log.p = FALSE then we are working with the probability of p.
      lower.tail  takes the logical value and lower.tail = TRUE by default.
                If it is true, then it indicates that we are calculating
                probabilities of P(X ≤ x).If lower.tail = FALSE, it indi-
                cates we are calculating the probability of P(X > x).
```

The lower tail of the normal curve is the direction of the tail to the left of a given point (value) while the upper tail is the direction to the right side of the curve at a given point. Thus, the lower-tailed test is when the critical region is on the left side of the distribution of the test value. The upper-tailed test is when the critical region is on the right side of the distribution of the test value. The *critical region* is denoted as α and represents the area within a sampling distribution that will lead to rejection of the hypothesis being tested when, in fact, the hypothesis is true.

The followings scripts are the solutions to Example 6.2, Example 6.3, and Example 6.4 above, with the use of R code. We get the same results in R ... with less stress.

Solution to Example 6.2 using R

```
# To calculate the probability of students scoring above 70
prob70<-pnorm (70, mean=65, sd=12.5, lower.tail=FALSE)
prob70
```

```
[1] 0.3445783
percent70<- prob70*100
percent70
[1] 34.45783
```

Solution to Example 6.3 using R

```
# (a) To calculate the probability that the machine will produce between 78 and
85?
btw78_85 <- pnorm(85, 80, 5) - pnorm(78, 80, 5)
btw78_85
[1] 0.4967665

# (b). To calculate the probability that the machine will produce fewer than 90?
less90 <- pnorm(90, 80, 5)
less90
[1] 0.9772499
```

Solution to Example 6.4 using R

```
# (a) To calculate the probability that new company registration will be fewer
than 35
prob35<-pnorm (35, mean=50, sd=4, lower.tail=FALSE)
prob35
[1] 0.9999116

# (b) To calculate the probability of new company registration will be exactly 45
prob45<- pnorm (45, mean=50, sd=4, lower.tail=TRUE)
prob45
[1] 0.1056498

# (c) To calculate the probability that new company registration will fall be-
tween 35 and 45
btw35_45 <- pnorm(45, 50, 4) - pnorm(35, 50, 4)
[1] 0.1055614
```

Exercises for Chapter 6

1. Suppose $X \sim Bin(100, 0.8)$. Calculate:
    (a) $P(x \leq 70)$
    (b) $P(x > 75)$

(c) $P(70 \le x \le 90)$

2. A machine packs rice in 50kg bags. It is observed that there is a variation in the actual weight that is normally distributed. Records show that the standard deviation of the distribution is 0.10 kg and the probability that the bag is underweight is 0.05. Find:
   (a) The mean value of the distribution.
   (b) What value standard deviation is needed to ensure that the probability that a bag is underweight is 0.005? Assume the mean is constant and improvement in the machine relies on reduction in the value of the standard deviation.

3. The Ozone cinemas revealed that movie customers spent an average of $20 on concessions with a standard deviation of $2. If the spending on concessions follows a normal distribution:
   (a) Find the percentage of customers that will spend less than $18 on concessions.
   (b) Find the percentage of customers that will spend more than $18 on concessions.
   (c) Find the percentage of customers that will spend more than $20 on concessions.

4. Suppose we know that the survival rate of a new business is normally distributed with mean 60% and standard deviation 8%.
   (a) What is the probability that more than 25 new businesses out of 30 new businesses registered in a week will survive?
   (b) What is the probability that between 20 and 25 new businesses will survive?

5. The average lifetime of a light bulb is 4500 hours with a standard deviation of 500 hours. Assuming that the average lifetime of light bulbs is normally distributed.
   (a) Find the probability that the average life of a bulb is between 4200 and 4700 hours.
   (b) Find the probability that the average life of a bulb exceeds 4500 hours.

6. Suppose the time spent (in minutes) in opening a new bank account is X with a probability density function of:

$$f(x) = \begin{cases} kx^2 \text{ for } x > 0 \\ 0 \text{ otherwise} \end{cases}$$

1. Find the value of k.
2. Find the probability that the time spent to open a new account will be less than 10 minutes.

7 Other Continuous Probability Distributions

This chapter explains some commonly used distributions in the tests of hypotheses. We will be focused on the formulas and their uses in statistics. These distributions are the Student's t-distribution, chi-square distribution, and F-distribution. We will see the conditions for using each of the distributions mentioned.

7.1 Student's t-Distribution

The t distribution was first discovered by W.S. Gosset and he published it under the pseudonym Student. The t distribution is, therefore, often called the Student's t-distribution.

Suppose we are to draw a random sample of n observations from a normally distributed population, then $z = \frac{\bar{x} - \mu}{\sigma/\sqrt{n}}$ has the standard normal distribution.

In most cases, we don't known the population standard deviation (σ) and we need to estimate it through the sample standard deviation(s). The basic fundamental is that the population standard deviation (σ) is a constant while sample standard deviation (s) varies from sample to sample. Thus, the quantity $\frac{\bar{x} - \mu}{s/\sqrt{n}}$ no longer has a standard normal distribution but a t-distribution $t = \frac{\bar{x} - \mu}{s/\sqrt{n}}$ with (n-1) degree of freedom.

Mathematically, it is written as, $t = \frac{\bar{x} - \mu}{s/\sqrt{n}} \sim t_{n-1}$.

This degree of freedom is obtained from the denominator of the sample variance, $\frac{s^2}{n-1}$. So the t-distribution and sample variance are highly connected. If we look at the quantity $\left(\frac{\bar{x} - \mu}{s/\sqrt{n}}\right)$, it looks very much like the z statistic which has normal distribution except when we replace the population standard deviation (σ) with (s). Then it has a greater variability, so t statistics is going to be similar to a normal distribution but with greater variance. The standard normal distribution and t-distribution have the mean = 0 at the center, but the difference is that the t-distribution has *heavier* tails (more area in the tails). As the degree of freedom increases, the t-distribution becomes the standard normal distribution.

In statistical inference, suppose we want to construct a 95% confidence interval for the population mean. Let's consider two cases when the population standard deviation is known and when population standard deviation is unknown.

https://doi.org/10.1515/9781547401475-007

When the population standard deviation (σ) is known, then we have a 95% confidence interval for the population mean when:

$$\bar{X} \pm 1.96 \times \frac{\sigma}{\sqrt{n}}$$

However, when the population standard deviation (σ) is unknown, we replace σ with s. The value 1.96 obtained from the standard normal distribution will no longer be appropriate, so look at t value that has the area of ($\alpha = \frac{0.05}{2} = 0.025$) to the right (i.e. $t_{0.025}$) because t distribution is of greater area in the tail and greater variability in the standard normal distribution, so we are expecting the $t_{0.025} > 1.96$.

How much greater can only be determined by the degree of freedom. Table 7.1 shows the value of the t-distribution at 95% for various sample sizes and degrees of freedom. You will notice that as the sample size increases the t-value decreases. When the degree of freedom (df) = ∞, then the t-value tends to standard normal distribution with the t-value of 1.96; this can also be seen in Figure 7.1.

Table 7.1: t-distribution extract.

Sample size	Degree of freedom	t-value$t_{0.025}$
6	5	2.571
11	10	2.228
21	20	2.086
41	40	2.021
81	80	1.99
∞	∞	1.96

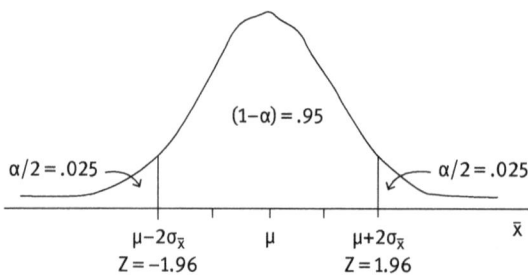

Figure 7.1: Standard normal curve showing area bounded for 95% confidence interval.

In general, when a sample size is less than 30 and the population standard deviation is unknown, then the random variable x is approximately normally distributed. Thus it follows that a Student's t-distribution is a suitable test. Furthermore, the difference between a t-distribution and normal distribution is negligible when the sample size is sufficiently large. When the population standard deviation is unknown, estimating the population standard deviation by using the sample standard deviation gives rise to greater uncertainty and a more spread-out distribution, i.e., heavier tails (see Figure 7.2).

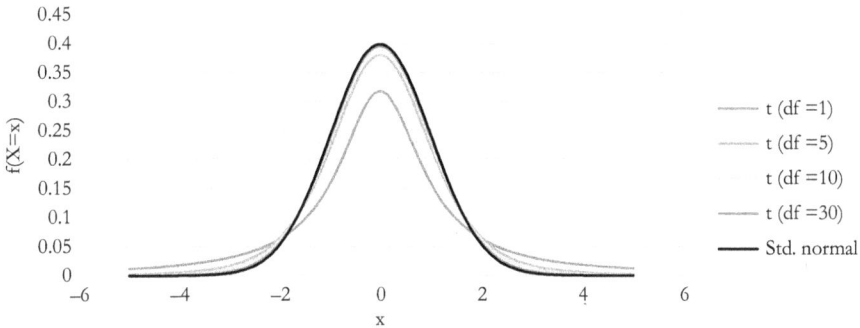

Figure 7.2: t-distribution with different degrees of freedom and standard normal distribution.

From Figure 7.2, we see that the graphs of the t-distribution are similar to a standard normal distribution except that a t-distribution is lower and wider; this attribute is prominent in the t-distribution with degree of freedom = 1. The graphs also show the absolute and relative error for normal approximation. The graph of a t-distribution with degree of freedom = 30 is approximately a standard normal distribution. Therefore, as the degrees of freedom increase, the t-distribution approaches the standard normal distribution.

When the population standard deviation is unknown, the sample size is less than 30, and the random variable x is approximately normally distributed, it follows a t-distribution. The test statistic is:

$$t = \frac{(\bar{x} - \mu)}{s/\sqrt{n}} \sim t_{(n-1),(\alpha/2)}$$

When you use a t-distribution to estimate a population mean, the degrees of freedom are equal to one less than the sample size: df = $n - 1$.

7.1.1 Properties of the Student's t-Distribution

The t-distribution has the following properties:
- The t-distribution is bell shaped and symmetric about the mean.
- The t-distribution is a family of curves, each determined by the degrees of freedom.
- The total area under a t-curve is 1 or 100%.
- The mean, median, and mode of the t-distribution are equal to zero.
- The variance of the t-distribution is $v/(v-2)$, where v is the degree of freedom and $v > 2$.
- When the degrees of freedom of a t-distribution is suffiently large, the t-distribution approaches the normal distribution.

Example 7.1:
What is the value of t in the following:
(i) $t_{0.05}(20)$
(ii) $t_{0.01}(20)$
(iii) $t_{0.975}(25)$
(iv) $t_{0.95}(25)$

Solution
Let's make use of the t-distribution table (see the appendix);

(i) $t_{0.05}(20) = 1.725$

We look at the t-distribution table where significance level (α) is 0.05 and degree of freedom (v) is 20, then we get 1.725.

The value (1.725) is the point at which two standard deviations from the normal is reached.

(ii) $t_{0.01}(20) = 2.528$

We look at the t-distribution table where significance level (α) is 0.01 and degree of freedom (v) is 20, then we get 2.528, this is the point at which three standard deviations from the normal reached.

(iii) $t_{0.975}(25) = 2.060$

Because of the symmetric nature of a t-distribution, when we look at the t-distribution table where significance level (α) is 0.025 and degree of freedom (v) is 25, then we get 2.060.

(iv) $t_{0.95}(25) = 1.708$

We look at the t-distribution table where significance level (α) is 0.05 and degree of freedom (v) is 25, and since t distribution is symmetric, we get 1.708.

7.1.2 R Code to Obtain t-values Using Example 7.1

```
The computation of the t-values in R is of the form:
qt(p, df, ncp, lower.tail = TRUE, log.p = FALSE)
where   qt is the quantile function for the t distribution
        p is the vector of probabilities
        ncp is the non-centrality parameter delta; if omitted, use the central
        t-distribution
        lower.tail  is logical; if TRUE (default), probabilities are P[X ≤ x],
        otherwise, P[X > x]
        log.p is logical; if TRUE, probabilities p are given as log(p)
```

For Example 7.1(i)-(iv), we can simply use the following R code:

Example 7.1(i)

```
alpha <- 0.05
df<-20t.alpha <- qt(alpha, df) t.alpha
[1] -1.724718
t.abs<-abs(t.alpha) # t-distribution is symmetric
t.abs
[1] 1.724718
```

Example 7.1(ii)

```
alpha<- 0.01
df<-20t.alpha <- qt(alpha, df, lower.tail =TRUE) t.alpha
[1] -2.527977
t.abs<-abs(t.alpha) # t-distribution is symmetric
t.abs
[1] 2.527977
```

Example 7.1(iii)

```
alpha<- 0.975
df<-25t.alpha <- qt(alpha, df, lower.tail =TRUE) t.alpha
[1] 2.059539
```

Example 7.1(iv)

```
alpha<- 0.95
df<-25t.alpha <- qt(alpha, df, lower.tail =TRUE) t.alpha
[1] 1.708141
```

Example 7.2:

Suppose a random number X follows a t-distribution with $v = 10$ degrees of freedom. Calculate the probability that the absolute value of X is less than 2.228.

Solution

$$P(|X| < 2.228) = P(-2.228 < X < 2.228)$$

We write the probability on the RHS in terms of cumulative probabilities:

$$P(|X| < 2.228) = P(X < 2.228) - P(X < -2.228)$$

Since t-distribution is symmetric and the t-table does not contain negative t-values, we can write the probabilities as:

$$P(|X| < 2.228) = P(X < 2.228) - P(X > 2.228)$$

From the t-table, the $P(X < 2.228) = 0.975$ and $P(X > 2.228) = 0.025$
The resulting probability is:

$$P(|X| < 2.228) = 0.975 - 0.025 = 0.95$$

The value of $|X| < 2.228$ falls on the two standard deviation of the normal curve (Figure 7.3).

7.2 Chi-square Distribution

In this subsection, we look at the definition, properties, and uses of the *chi-square* test. Also referred to as chi-square, it is denoted by $\chi^2(v)$ with degree of freedom (v). Chi-square is useful mainly when the outcome is discrete (dichotomous, ordinal, or categorical) data. The chi-square statistic makes a comparison between the observed count in the contigency table and the expected count under the assumption that there is no association between row and column classifications. Therefore, the chi-square statistic is used in hypothesis testing of no association between two or more groups. For example, a study wants to investigate the effectiveness of a new employment handbook. Out of 100

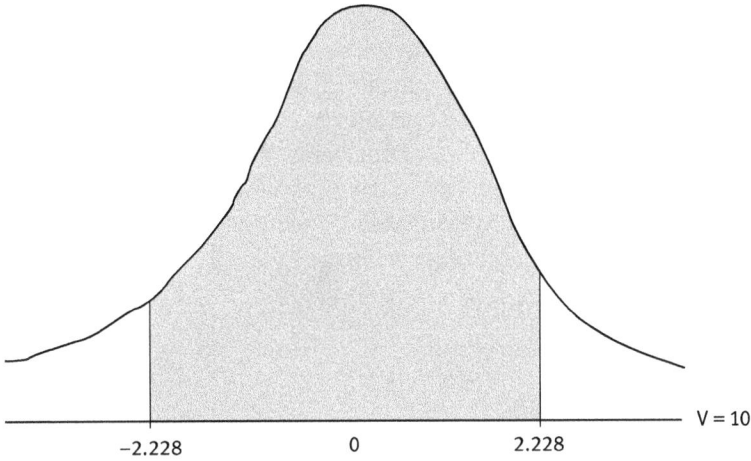

Figure 7.3: Normal curve for the absolute value of X is less than 2.228.

employees sampled, composed of 65 are men and 35 are women, 50 men and 30 women respond positively to the handbook. We may want to determine whether the response to the handbook depends on gender; with the use of the chi-square test we can determine whether the response to a handbook has this association.

The contingency table for this information is shown in Table 7.2.

Table 7.2: Contingency table (2×2) for medication survey.

Response	Gender	
	Male	**Female**
Positive	50	30
Negative	15	5
Total	65	35

Let Y_1, Y_2, ..., Y_n be mutually independent standard normal random variables, and let $X = Y_1^2 + Y_1^2 + \ldots + Y_n^2$ have a chi-square distribution. The probability distribution function of the chi-square distribution is:

$$f(x) = \frac{e^{-x/2}x^{v/2}}{2^{v/2}\Gamma\left(\frac{v}{2}\right)}, \quad \text{for } x \geq 0$$

Where v is the shape parameter (df) and Γ is the gamma function. For instance, $\Gamma(x) = (x-1)!$

Since the chi-square distribution is a square of standard normal, the random variable takes only non-negative values and it tends to be right skewed. Its skewness depends on the degrees of freedom or number of observations. As the number of observations increases, the more symmetrical the chi-square distribution becomes, and then tends to normal distribution. The higher the degrees of freedom, the less the skewness of the chi-square (Figure 7.4).

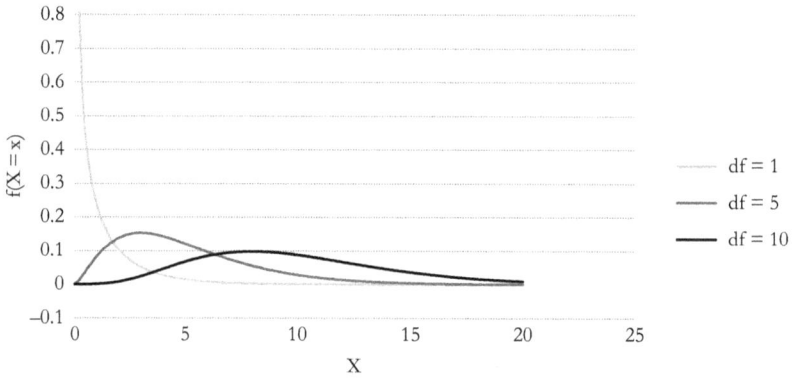

Figure 7.4: Chi-square distribution with different degrees of freedom.

The chi-squared test shown mathematically is:

$$X^2 = \sum_{i-1}^{r} \sum_{j=1}^{c} \frac{\left(O_{ij} - E_{ij}\right)^2}{E_{ij}} \sim X_v^2$$

where O_{ij} is the observed value, E_{ij} is the expected value, and v is the degree of freedom.

The expected values can be computed as follows:

$$E_{ij} = \frac{\sum_{i=1}^{r} O_{ij} \times \sum_{i=1}^{C} O_{ij}}{N}$$

$$E_{ij} = \frac{sum\ of\ ith\ row \times sum\ of\ jth\ column}{N}$$

where N is the total observations.

Example 7.3:
Use the information in Table 7.2 to test whether there is any gender bias in response to the new employment handbook.

Solution:
Given that $r = 2$, $c = 2$, df $= (r-1)(c-1) = 1$ and $\alpha = 0.05$

Hypothesis:
H_0: There is no gender bias in the response to the new employment handbook
H_1: There is gender bias in the response to the new employment handbook
Test statistic:

$$\chi^2 = \sum_{i=1}^{r} \sum_{j=1}^{c} \frac{(O_{ij} - E_{ij})^2}{E_{ij}}$$

The expected values are computed as follows:

$$\chi^2 = \frac{(O_{ij} - E_{ij})^2}{E_{ij}}$$

$$E_{11} = \frac{80 \times 65}{100} = 52, E_{12} = \frac{80 \times 35}{100} = 28, E_{21} = \frac{20 \times 65}{100} = 13, E_{22} = \frac{20 \times 35}{100} = 7$$

$$\chi^2 = \frac{(50-52)^2}{52} + \frac{(30-28)^2}{28} + \frac{(15-13)^2}{13} + \frac{(5-7)^2}{7} = 1.10$$

Critical value:
$\chi^2_{\alpha=0.05}(1) = 3.841$

Decision rule:
Reject H_0 if $\chi^2 > \chi^2_{\alpha=0.05}(1)$. Since $1.10 < 3.841$, then, we do not reject null hypothesis.

Conclusion:
There is no gender bias in the response to the new employment handbook.

7.2.1 Properties of the Chi-square Distribution

The chi-square distribution has the following properties:
- The chi-square distribution is a continuous probability distribution with values ranging from 0 to ∞ (nonnegative).
- The mean of a chi-square distribution is the number of degrees of freedom, v.
- The variance of a chi-square distribution is the number of degree of freedom doubled, $2v$.
- Chi-square has an additive property. The sum of independent χ^2 is itself a χ^2 variate. Suppose two independent χ_1^2 and χ_2^2 with the degrees of freedom of v_1 and v_2 respectively, then $\chi_1^2 + \chi_2^2$ is the same as χ^2 variate with degree of freedom $v_1 + v_2$.

7.2.2 Uses of the Chi-square Distribution

Chi-square can be used in the following ways:

1. The chi-square tests two or more independent comparison groups and the outcome of interest is discrete with two or more responses. The responses can be dichotomous, ordinal, or categorical. For two or more independent comparison groups, we can compare the distribution of responses to the discrete outcome variable among several independent comparison groups under the null hypothesis. The null hypothesis states that there is no difference in the distribution of responses to the outcome across comparison groups.

2. The chi-square test is used to determine whether there is association between two categorical variables in a sample which is likely to reflect a real association between these two variables in the population. The information may be better presented in a contigency table to test if there is association between the row group and the columns group.

3. The chi-square test is used to compare the observed distribution to an expected distribution. That is, the closeness of the observed values to those which would be expected under the fitted model. The null and alternative hypotheses for our *goodness of fit* test reflect the assumption we are making about the population regarding consumers' behavior, consumers' acceptance and preference for a new product etc.

Note that the t-distribution is the ratio of the standard normal distribution and square root of chi-square divided by its degree of freedom (df). The degrees of freedom are the number of free choices left after removing a sample statistic.

Example 7.4:

What is the value of chi-square in the following:

(i) $\chi^2_{0.05}(15)$

(ii) $\chi^2_{0.995}(30)$

(iii) $\chi^2_{0.900}(85)$

Solution

We look at the chi-square table for the significance level (α) and the degree of freedom (v).

(i) $\chi^2_{0.05}(15) = 7.261$

(ii) $\chi^2_{0.995}(30) = 53.67$

(iii) $\chi^2_{0.900}(85) = 102.07$

We got the answer for $\chi^2_{0.900}(85) = 68.7845$ from finding the average of $\chi^2_{0.900}(80) = 96.5782$ and $\chi^2_{0.900}(90) = 107.565$, since we can not see $\chi^2_{0.900}(85)$ directly from the chi-square table, we use the average method to estimate the value.

Note that the chi-square table has the probabilities of P[X ≤ x]. That is we consider the value of x from the left hand side of the curve as shown in Figure 7.5.

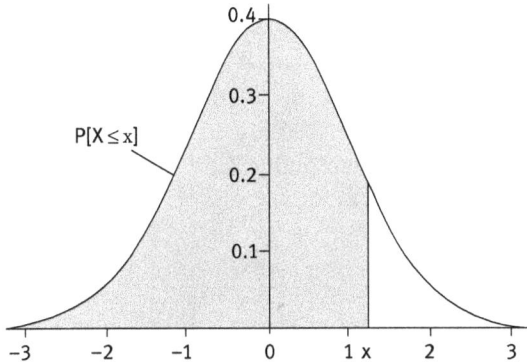

Figure 7.5: Chi-square with probability P[X ≤ x].

R code to obtain chi-square values from the table using Example 7.4
The chi-square can be obtained in R using the function:

```
qchisq(p, df, ncp, lower.tail = TRUE, log.p = FALSE)
where qchisq is the quantile function for the Chi-square distribution
        p is the vector of probabilities
        ncp is the non-centrality parameter delta; if omitted, use the central χ²
        distribution
        lower.tail is logical; if TRUE (default), probabilities are P[X ≤ x],
        otherwise, P[X > x]
        log.p is logical; if TRUE, probabilities p are given as log(p)
```

The following are the R code to solve the Example 7.4(i)-(iii):

```
qchisq(0.05, df=15, lower.tail = TRUE)
[1] 7.260944

qchisq(0.995, df=30, lower.tail = TRUE)
[1] 53.67196

qchisq(0.900, df=85, lower.tail = TRUE)
[1] 102.0789
```

We easily got the inverse of the p-value for the chi-square distribution in Example 7.4.

7.3 F-Distribution

The *F-distribution* is the ratio of two independent chi-square distributions divided by their degrees of freedom v_1 and v_2. This simply means $\frac{chi-square\ A}{v_1} \div \frac{chi-square\ B}{v_2}$

The F-distribution is mainly used when we are working with ratios of variances.

The probability density function of the F-distribution is:

$$f(x) = \frac{\Gamma\left(\frac{v_1+v_2}{2}\right)\left(\frac{v_1}{v_2}\right)^{\frac{v_1}{2}} x^{\frac{v_1}{2}-1}}{\Gamma\left(\frac{v_1}{2}\right)\Gamma\left(\frac{v_2}{2}\right)\left(1+\frac{v_1 x}{v_2}\right)^{\left(\frac{v_1+v_2}{2}\right)}}, \quad x \geq 0$$

where v_1 and v_2 are the shape parameters (the degrees of freedom); these degrees of freedom are positive integers and Γ is the gamma function.

The F-distribution is mainly spread out when the degrees of freedom are small. Thus, as the degrees of freedom decrease, the F-distribution is more dispersed. The F-distribution is asymmetric and has a minimum value of 0 and maximum of infinity. The curve reaches a peak not far to the right of 0, and then gradually approaches the highest value of F on the horizontal axis. The F-distribution approaches but never touches the horizontal axis.

7.3.1 Properties of the F-Distribution

The F-distribution has the following properties:
- The F-distribution is asymmetric about the mean but skewed to the right.
- The F-distribution is a family of curves that are determined by the two degrees of freedom.
- The F-distribution is a continuous probability distribution with the values ranging from 0 to ∞ (nonnegative).
- The curve approaches normal as the degrees of freedom for the numerator and denominator become larger.
- The mean of the F-distribution is $\frac{v_2}{v_2-2}$
- The variance of the F-distribution is $\frac{2v_2^2(v_1+v_2-2)}{v_1(v_2-2)^2(v_2-4)}$

– Any changes in the first parameter (v_1) does not change the mean of the distribution, but the density of the distribution is shifted from the tail of the distribution toward the center of the curve. However, the F-distribution remains asymmetric.

7.3.2 Uses of the F-Distribution

The F-distribution is useful in the following:

1. The F-distribution is used to test the validity of a multiple regression model. For a multiple regression model with intercept, we may want to test the null hypothesis and alternative hypothesis:

$$H_0: \beta_1 = \beta_2 = ... = \beta_{p-1} = 0$$

$$H_1: \beta_1 \neq \beta_2 \neq ... \neq \beta_{p-1} \neq 0$$

So, the F test is used to test the significance of overall regression coefficients in a multiple regression model.

2. The F-distribution is used to test hypotheses about the equality of two independent population variances. If s_1^2 and s_2^2 are sample variances from two independent populations, the ratio of the two sample variances is F, that is $F = \frac{s_1^2}{s_2^2}$. Therefore, we assume that the ratio of variances is 1 when the two variances are equal. Consequently, our null hypothesis states that the population variances are equal.

3. It is used to draw an inference about data that is drawn from a population in analysis of variance (ANOVA).

Example 7.5:
What is the value of F in the following:
(i) $F_{(0.05)}(8, 12)$
(ii) $F_{(0.95)}(10, 15)$
(iii) $F_{(0.995)}(20, 50)$

Solution
Let's look at the F table (see Appendix) for significance level (α) and degrees of freedom $(v_1$ and $v_2)$ for the numerator and denominator respectively.
(i) $F_{(0.05)}(8, 12) = 0.3045$
(ii) $F_{(0.95)}(10, 15) = 2.544$
(iii) $F_{(0.995)}(20, 50) = 2.470$

Note that we use the F table that has the probabilities of P[X ≤ x], that is the curve starts the reading from the left side of the curve (see Figure 7.6).

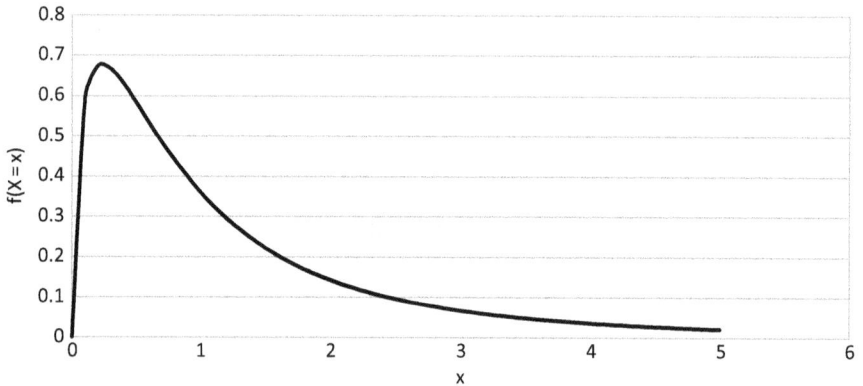

Figure 7.6: F-distribution with degrees of freedom ($v_1 = 3$ *and* $v_2 = 5$).

R Code to Obtain F Values from the F-Distribution Table Using Example 7.5
The F value can be obtained in R using the function:

```
qf(p, df1,df2, ncp, lower.tail = TRUE, log.p = FALSE)
```

where qf is the quantile function for the F distribution
  p is the vector of probabilities
  df1 is the degree of freedom for the numerator
  df2 is the degree of freedom for the denominator
  ncp is the non-centrality parameter delta; if omitted, use the central F-distribution
  lower.tail is logical; if TRUE (default), probabilities are P[X ≤ x], otherwise, P[X > x]
  log.p is logical; if TRUE, probabilities p are given as log(p)

With these short R statements we are able to obtain the F values for Example 7.5(i)-(iii).

```
qf (0.05, 8, 12, lower.tail = TRUE, log.p = FALSE)
[1] 0.3045124

qf (0.95, 10, 15, lower.tail = TRUE, log.p = FALSE)
[1] 2.543719

qf (0.995, 20, 50, lower.tail = TRUE, log.p = FALSE)
[1] 2.470161
```

Exercises for Chapter 7

1.
   a. Mention a continuous probability distribution you know and what it is used for.
   b. Define Student's t-distribution and state its properties.
   c. Explain how a t-distribution can be used as an approximation of the normal distribution.
   d. What is the value of t in the following: (i) $t_{0.95}(18)$, (ii) $t_{0.1}(25)$, (iii) $t_{0.995}(30)$, and (iv) $t_{0.99}(10)$?

2.
   a. Define the chi-square distribution and state its properties.
   b. What are the uses of the chi-square distribution?
   c. What is the value of chi-square in the following: (i) $\chi^2_{0.95}(20)$, (ii) $\chi^2_{0.995}(40)$, (iii) $\chi^2_{0.975}(25)$, and (iv) $\chi^2_{0.99}(50)$?

3.
   a. Define the F-distribution and state its properties.
   b. What are the uses of the F-distribution?
   c. What is the value of F in the following: (i) $F_{(0.95)}(10,12)$, (ii) $F_{(0.99)}(5,25)$, (iii) $F_{(0.05)}(10,30)$, and $F_{(0.01)}(18,10)$?

4. A marketing department administered 5000 questionaires to 3 age categories consumers to investigate their preference in the 2 different types of a product. The table below shows the outcome of the survey.

Age Categories	Type A	Type B	Total
Teenagers	1300	1200	2500
Youths	625	875	1500
Old	435	565	1000
Total	2360	2640	5000

Test whether the age category influenced the preference among different types of the product.

8 Sampling and Sampling Distribution

Sampling distributions are derived from a group of selected data being calculated using statistics such as mean, median, mode, range, standard deviation, and variance. These distributions are useful in testing hypotheses and making inferences about a population.

The sampling distribution relies on the following:
- An underlying distribution of the population
- The statistic under investigation
- Sampling techniques
- Sample size

In this chapter, we consider probability and non-probability sampling; sampling techniques; sampling distribution of the mean; and proportion and the central limit theorem.

8.1 Probability and Non-Probability Sampling

A method that uses random sampling techniques to generate a sample where every sampling unit in the population has an equal chance of being selected is called a *probability sampling*. Randomization is the key in probability sampling techniques and is designed to ensure that all individuals have equal probability of selection. Suppose a population contains 100 sampling units, each sampling unit has a 1/100 chance of being selected by probability sampling techniques. Under probability sampling, the sample selected gives a true representation of the population that it comes from. Most common probability samplings are simple random, systematic, stratified, and cluster samples.

Advantages of probability sampling:
- Outcomes are reliable
- Increases accuracy of error estimation
- No sampling bias and systematic error
- Good for making inferences about the population

Disadvantages of probability sampling:
- Time consuming
- More expensive than non-probability sampling
- More complex than non-probability sampling

https://doi.org/10.1515/9781547401475-008

Non-probability sampling involves non-random sampling of the sampling units, thus only specific members of the population have a chance of being selected. The most widely used non-probability methods are judgment sampling, quota sampling, convenience sampling, and extensive sampling. Non-probability sampling techniques are based on subjective judgment and are used for exploratory studies (e.g., a pilot (feasibility survey).

    Advantages of non-probability sampling:
- Easily generates a description of the sample
- Saves time
- Less costly compared to probability sampling
- Can be used when it is not feasible to conduct probability sampling

Disadvantages of non-probability sampling:
- Lack of representation of the population
- Difficult to estimate sampling variability
- Difficult to identify possible selection bias
- Lower level of generalization of research findings

8.2 Probability Sampling Techniques—Simple Random, Systematic, Cluster, and Stratified

8.2.1 Simple Random Sampling (SRS)

In a *simple random sampling* technique, each unit included in the sample has an equal chance of inclusion in the sample. In a homogenous population, simple random sampling provides an unbiased and better estimate of the population parameters. In the process of selection, the probability of selecting the first sampling unit will not neccessarily be the same probability as selecting the second sampling unit.

    In the event that every selection of a sampling unit from a population remains constant with a chance of $1/N$, where N is the total number of elements in a population, then the procedure is called *simple random sampling with replacement* (SRSWR). In a SRSWR, there are N^n possible samples with each sample having a $\dfrac{1}{N^n}$ chance of selection, where n is the sample size.

    If in the first selection each member of the population has an equal chance of selection, with probability of $1/N$; and in the selection of second unit the first unit is excluded, all the remaining sampling units have probability of $\dfrac{1}{N-1}$ of selection. Any successive selection in an SRS is referred as to a *simple random sample without replacement* (SRSWOR). If sampling is done without replacement, there are $\binom{N}{n}$ possible samples and each sample has a $\dfrac{1}{\binom{N}{n}}$ chance of selection. More often than not, the SRSWOR is simply regarded as a *simple random sample*.

The following methods are used to select units of a simple random sample:

1. Random Number Table

In using a random number table (Table 8.1), all units of the population are numbered either from 1 to N or 0 to N-1. Suppose we have a population size of 50 and we want to select a sample size of 10. We can number sampling units in

Table 8.1: Random digits.

11164	36318	75061	37674	26320	75100	10431	20418	19228	91792
21215	91791	76831	58678	87054	31687	93205	43685	19732	08468
10438	44482	66558	37649	08882	90870	12462	41810	01806	02977
36792	26236	33266	66583	60881	97395	20461	36742	02852	50564
73944	04773	12032	51414	82384	38370	00249	80709	72605	67497
49563	12872	14063	93104	78483	72717	68714	18048	25005	04151
64208	48237	41701	73117	33242	42314	83049	21933	92813	04763
51486	72875	38605	29341	80749	80151	33835	52602	79147	08868
99756	26360	64516	17971	48478	09610	04638	17141	09227	10606
71325	55217	13015	72907	00431	45117	33827	92873	02953	85474
65285	97198	12138	53010	94601	15838	16805	61004	43516	17020
17264	57327	38224	29301	31381	38109	34976	65692	98566	29550
95639	99754	31199	92558	68368	04985	51092	37780	40261	14479
61555	76404	86210	11808	12841	45147	97438	60022	12645	62000
78137	98768	04689	87130	79225	08153	84967	64539	79493	74917
62490	99214	84987	28759	19177	14733	24550	28067	68894	38490
24216	63444	21283	07044	92729	37284	13211	37485	10415	36457
16975	95428	33226	55903	31605	43817	22250	03918	46999	98501
59138	39542	71168	57609	91510	77904	74244	50940	31553	62562
29478	59652	50414	31966	87912	87154	12944	49862	96566	48825
96155	95009	27429	72918	08457	78134	48407	26061	58754	05326

Table 8.1 (continued)

29621	66583	62966	12468	20245	14015	04014	35713	03980	03024
12639	75291	71020	17265	41598	64074	64629	63293	53307	48766
14544	37134	54714	02401	63228	26831	19386	15457	17999	18306
83403	88827	09834	11333	68431	31706	26652	04711	34593	22561
67642	05204	30697	44806	96989	68403	85621	45556	35434	09532
64041	99011	14610	40273	09482	62864	01573	82274	81446	32477
17048	94523	97444	59904	16936	39384	97551	09620	63932	03091
93039	89416	52795	10631	09728	68202	20963	02477	55494	39563
82244	34392	96607	17220	51984	10753	76272	50985	97593	34320

two-digits, such as 01, 02, ..., 49, 50. We read a two-digit random number from the random number table, any two-digit numbers that are greater than 50 are ignored and any repetition of a number is also ignored if the sampling is without replacement.

We can read a random number from any column or row of the random number table. From Table 8.1, let's start from first column downward to form our two-digit numbers. In the left column we have: 11, 21, 10, 36, 49, 17, 24, 16, 29, and 12. Some numbers such as 73, 64, 51, 99, 71, 65, 95, 61, 78, 62, 59, 96 are discarded from the list because they are not part of the population. Furthermore, let's consider a population of 100, and we are asked to take a random sample of 10 from the population. We read the random number table in 3-digits—that is, we will assign a number from 001 to 100. From Table 8.1 we have: 047, 046, 098, 070, 024, 088, 004, 084, 094, and 097 in that order.

2. Lottery Method

The lottery method is very close to modern methods of sample selection. This process involves labeling all sampling units from 1 to N in the sampling frame. The same numbers (1 to N) are written on slips of paper, then the slips are mixed thoroughly and selections are made randomly. This method is advisable when the size of the population is small. It may be impractical if the sample size is large since it is not easy to properly mix a heap of paper slips.

1. Computer Generation

Computers have built-in programs that help to generate random samples with ease. These are mostly used to select of winners of applicants for plots of land, visa lotteries, and so on.

Example 8.1:
Suppose a population contains the following: 4, 6, 10, 11, 15, 17, and 20 units. To select a sample size of 2, what are the possible samples (a) with replacement and (b) without replacement?

Solution
(a) SRSWR:
Possible samples = $7^2 = 49$
The lists of possible samples for SRSWR are displayed in Table 8.2.

Table 8.2: Possible samples for SRSWR, Example 8.1.

Samples	4	6	10	11	15	17	20
4	(4, 4)	(4, 6)	(4, 10)	(4, 11)	(4, 15)	(4, 17)	(4, 20)
6	(6, 4)	(6, 6)	(6, 10)	(6, 11)	(6, 15)	(6, 17)	(6, 20)
10	(10, 4)	(10, 6)	(10, 10)	(10, 11)	(10, 15)	(10, 17)	(10, 20)
11	(11, 4)	(11, 6)	(11, 10)	(11, 11)	(11, 15)	(11, 17)	(11, 20)
15	(15, 4)	(15, 6)	(15, 10)	(15, 11)	(15, 15)	(15, 17)	(15, 20)
17	(17, 4)	(17, 6)	(17, 10)	(17, 11)	(17, 15)	(17, 17)	(17, 20)
20	(20, 4)	(20, 6)	(20, 10)	(20, 11)	(20, 15)	(20, 17)	(20, 20)

(b) SRSWOR:
Possible samples = $\binom{7}{2} = 21$
The lists of possible samples for SRSWOR are displayed in Table 8.3.

Table 8.3: Possible samples for SRSWOR, Example 8.1.

Samples	4	6	10	11	15	17	20
4							
6	(6, 4)						
10	(10, 4)	(10, 6)					
11	(11, 4)	(11, 6)	(11, 10)				
15	(15, 4)	(15, 6)	(15, 10)	(15, 11)			
17	(17, 4)	(17, 6)	(17, 10)	(17, 11)	(17, 15)		
20	(20, 4)	(20, 6)	(20, 10)	(20, 11)	(20, 15)	(20, 17)	

8.2.2 Systematic Sampling

Systematic sampling is easier than simple random sampling although it uses technique of SRS to select the first sample from a population. Each sampling unit has an equal chance of selection in the first sample taken from the population. The subsequent selections follow a predefined pattern. Consider a population of N from which a sample size of n is drawn. Assuming that $k = N/n$ initially, a random sample taken from 1 to k will be the first sample, then every kth unit in the population is taken from a sample systematically. For example, a recruiting agency received 1,000 qualified applicants, out of which only 50 applicants are to be considered for jobs. If they decide to interview only 50 applicants by systematic sampling, then $k = \dfrac{1000}{50} = 20$. Therefore, a random sampling shall be done between 1 to 20, say the 13th item is selected, so every 20th applicant in the population of 1,000 will form a sample (i.e., 13th, 33rd, 53rd, ..., 953rd, 973rd and 993rd).

8.2.2.1 Advantages of Systematic Sampling
– Simple and convenient to actualize
– Represents a population more easily without numbering each member of a sample
– Based on precision in member selection
– Minimal risk involved in the process of selection
– Can be used in a heterogenous population

8.2.2.2 Disadvantages of Systematic Sampling
– Cannot be used when the size of the population is not available or is being approximated
– Requires a population characterized with a natural degree of randomness in order to mitigate risk of selecting samples
– Higher risk of manipulation of data, increasing the likelihood of achieving a targeted outcome rather than randomness of a dataset

8.2.3 Cluster Sampling

Cluster sampling is a technique of selecting all individuals or a group of individuals through a random selection. Its procedure involves selection of groups—or *clusters*—then individual subjects are selected from each group by either SRS or systematic random sampling. It is desirable for clusters to be made up of heterogeneous units or elements. The samples can be an entire cluster or group rather than a subset of the cluster or group. For example, let's say the goal is to take a sample ofcompanies in the agricultural industries in six regions. It is fair to have sample representation from each of the regions rather than concentrating on 3 or 4 regions. The commonly used cluster here is geographical location.

Cluster sampling can either be single-stage or multi-stage. In a single-stage cluster sampling, clusters are selected by simple random sampling and then data is collected from every unit of the sampled clusters. A two-stage process involves a simple random sampling of clusters and then a simple random sampling of the units in each of the sampled clusters.

8.2.3.1 Advantages of Cluster Sampling
- Economical to generate
- Less time consuming for listing and implementation
- Works for a larger sample with a similar fixed cost

8.2.3.2 Disadvantages of Cluster Sampling
- Gives less information about each observation than SRS
- May show variation within a group
- Elements of the cluster may possess similar attributes
- High estimates of standard errors when compared with other probability sampling techniques

8.2.4 Stratified Random Sampling

Stratified random sampling is a probability sampling technique in which the population is partitioned into relatively homogeneous groups know as *strata*. In each stratum, a simple random sample is used to select sampling units. Thus, the statistic computed from the strata is aggregated to make inferences about the population. Suppose a team of researchers wants to compare productivity among five departments in a corporation (finance, IT, marketing, research & development, and legal). There are 50 employees in the finance department, 120 employees in IT department, 80 employees in marketing department, 15 employees in research & development department, and 35 employees in legal department. The distribution of the employee population is depicted in Figure 8.1.

If a stratified random sample of 30 employees are selected in the above example, we select employees according to the size of their departments (i.e., 5 employees from finance, 12 employees from IT, 8 employees from

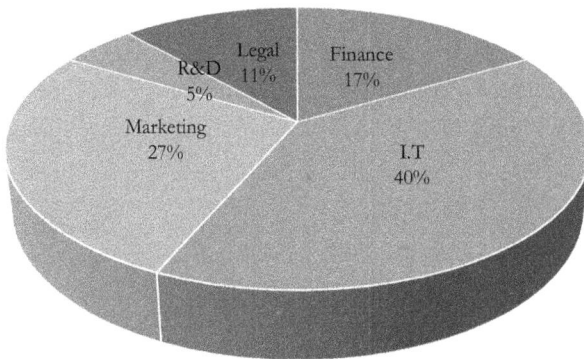

Figure 8.1: The distribution of employees by department.

marketing, 2 employees from research and development, and 4 employees from legal) in order to achieve better precision.

8.2.4.1 Advantages of Stratified Random Sampling
– High representation of the population
– Allows a valid statistical inference from the data collected
– Improves the potential for units to be evenly distributed over the population (hence, improves precision when compared with SRS)

8.2.4.2 Disadvantages of Stratified Random Sampling
– Cannot be practically used if a complete list of the population is not available
– The list of the population is needed to be clearly delineated into each stratum

8.3 Non-Probability Sampling Techniques—Judgment, Quota, and Convenience

8.3.1 Judgment Sampling

Judgment sampling is a type of non-probability sampling technique where researchers use their own experience and prior knowledge about a situation before commencement of sample selection—the sample is taken without using any statistical tools. Suppose that a panel wants to understand the reasons why start-up companies fail. The panel should form the samples from the business owners in order to obtain useful information about the factors responsible for such failures. This is the process we call judgment sampling.

8.3.1.1 Advantages of Judgment Sampling
- Easily understood
- Requires no specific knowledge of statistics
- Less time consuming to actualize

8.3.1.2 Disadvantages of Judgment Sampling
- The researcher may be biased in selecting the sample
- No logic in the process of sample selection
- No capacity to extrapolate data to the population since sample selection is not representative
- Can give rise to vague conclusions

8.3.2 Quota Sampling

Quota sampling is a non-probability sampling technique whereby the samples selected have the same proportions of individuals as the entire population in regard to known characteristics. Its procedures include:
- Dividing the population into exclusive sub-groups
- Identifying a portion of subgroups in the population to be maintained in the sampling process
- Selecting subjects from subgroups using the same proportion as in the previous step
- Ensuring the sample drawn is representative of the population, making it easy to understand the characteristics noticed in each subgroup. For example, if a population of 2000 units consists of 1200 men and 800 women and a researcher is interested in 30% of the population, a sample of 600 sampling units made up of 360 men and 240 women will be selected.

8.3.2.1 Advantages of Quota Sampling
- Allows sampling of a subgroup of interest and investigates the traits and characteristics
- Helps to establish relationships in terms of the traits and characteristics among subgroups

8.3.2.2 Disadvantages of Quota Sampling
- Selected characteristics of the population are accounted for while forming the subgroups
- The process may result in an over-represented sample

8.3.3 Convenience Sampling

The *convenience sampling* approach is an example of non-probability sampling where a researcher draws a sample because of the "convenience" of the data sources. Here, convenience could be the nearest subjects or the availability of resources for carrying out a research project. It is also known as *accidental sampling.* It is commonly used in pilot studies before embarking on the main project. For example, to determine the profitability of firms, a researcher may first look at some firms in the vicinity and sample opinions of the business owners in the area before extending the research work to larger communities.

8.3.3.1 Advantages of Convenience Sampling
–   Time efficient
–   Outcomes can be used to refine some questions in a questionnaire
–   Less costly, since a researcher is able to use available resources

8.3.3.2 Disadvantages of Convenience Sampling
–   Not a good representative of the entire population
–   May lead to a biased outcome and wrong conclusions because of uncontrollable sample representation

8.4 Sampling Distribution of the Mean

The mean of the sampling distribution is equal to the mean of the population, that is, $\mu_{\bar{x}} = \mu$.

To establish this fact, we can use the data in Example 8.1:

Suppose a sample size of 2 is drawn from a population that contains the following: 4, 6, 10, 11, 15, 17, and 20 units.

$$\text{The population mean} = \frac{4 + 6 + 10 + 11 + 15 + 17 + 20}{7} = 11.86$$

For SRSWR, the sample means are shown in Table 8.4.

$$\mu_{\bar{x}} = \frac{4.0 + 5.0 + 7.0 + \ldots + 17.5 + 18.5 + 20.0}{49} = 11.86$$

Therefore, the mean of distribution of sample means is the same as the population mean.

Table 8.4: The means of the possible samples for SRSWR.

Samples	4	6	10	11	15	17	20
4	4.0	5.0	7.0	7.5	9.5	10.5	12.0
6	5.0	6.0	8.0	8.5	10.5	11.5	13.0
10	7.0	8.0	10.0	10.5	12.5	13.5	15.0
11	7.5	8.5	10.5	11.0	13.0	14.0	15.5
15	9.5	10.5	12.5	13.0	15.0	16.0	17.5
17	10.5	11.5	13.5	14.0	16.0	17.0	18.5
20	12.0	13.0	15.0	15.5	17.5	18.5	20.0

For SRSWOR, the distribution of the sample means are shown in Table 8.5.

$$\mu_{\bar{X}} = \frac{5.0 + 7.0 + 7.5 + \ldots + 16.0 + 17.5 + 18.5}{21} = 11.86$$

Table 8.5: The means of the possible samples for SRSWOR.

Samples	4	6	10	11	15	17
20						
4						
6	5.0					
10	7.0	8.0				
11	7.5	8.5	10.5			
15	9.5	10.5	12.5	13.0		
17	10.5	11.5	13.5	14.0	16.0	
20	12.0	13.0	15.0	15.5	17.5	18.5

Hence, the mean of distribution of sample means is the same as the population mean.

In summary, irrespective of the sampling processes, the mean of distribution of means still remains the mean of the population.

The standard error of the mean is equal to the ratio of the standard error of the population to the square root of the sample size. This can also be demonstrated using the possible samples for SRSWR in Table 8.4.

The standard deviation of the population 4, 6, 10, 11, 15, 17, and 20 units is:

$$\sigma = \frac{(4-11.86)^2 + (6-11.86)^2 + \ldots + (20-11.86)^2}{7} = 5.38$$

Similarly, the standard deviation of the distribution of sample means is calculated as:

$$\sigma_{\bar{X}} = \frac{(4-11.86)^2 + (5-11.86)^2 + \ldots + (18.5-11.86)^2 + (20-11.86)^2}{49} = 3.8065$$

$$\sigma_{\bar{X}} = \frac{\sigma}{\sqrt{n}} = \frac{5.38}{\sqrt{2}} = 3.8065$$

This result shows the relationship between the standard deviation of distribution of means and the population standard deviation. That is, the standard deviation of distribution of means ($\sigma_{\bar{X}}$) is the population standard deviation (σ) divided by the square root of sample size (n).

Sampling distribution of the mean using R

```
# To compute the mean of the population
pop=c(4,6,10, 11, 15, 17, 20)
pop_mean<-sum(pop)/7
pop_mean
[1] 11.85714

# To compute the mean of the distribution of means for SRS with replacement
sWR<-49
sampWR<- c(4, 5, 7, 7.5, 9.5, 10.5, 12, 5, 6, 8, 8.5, 10.5, 11.5, 13, 7, 8, 10,
10.5, 12.5, 13.5, 15, 7.5, 8.5, 10.5, 11, 13, 14, 15.5, 9.5, 10.5, 12.5, 13, 15,
16, 17.5, 10.5, 11.5, 13.5, 14, 16, 17, 18.5, 12, 13, 15, 15.5, 17.5, 18.5, 20)
sampWR_mean=sum(sampWR)/sWR
sampWR_mean
[1] 11.85714

# To compute the standard deviation of the distribution of means for SRS with re-
placement
sampWR_std<-sd(sampWR)
sampWR_std
[1] 3.845994

# To compute the mean of the distribution of means for SRS without replacement
 sWOR<-21
sampWOR<-c(5,7,7.5,9.5,10.5,12,8,8.5,10.5,11.5,13,10.5,12.5,13.5,15,13,14,15.5,16,
17.5,18.5)
sampWOR_mean<-sum(sampWOR)/sWOR
sampWOR_mean
```

```
[1] 11.85714

# Population standard deviation
p<-7
pop_std<-sd(pop)*sqrt((p-1)/p)
pop_std
[1] 5.38327

# Sample standard deviation
n<-2   # sample selection of 2
samp_std<- pop_std/sqrt(n)
samp_std
[1] 3.806546
```

8.5 Central Limit Theorem and Its Significance

The *central limit theorem* states that if we let $X_1, X_2, \ldots, X_n$ be a random sample from a distribution with a finite mean (μ) and a finite variance (σ^2), for a sufficiently large sample size n (for example, ($n \geq 30$)), the following holds:

- The sample mean ($\bar{X}$) is approximately normal
- The mean of the sample means is equal to the population mean $E(\bar{X}) = \mu$
- Variance $var(\bar{X}) = \sigma_{\bar{X}}^2 = \dfrac{\sigma^2}{n}$

This theorem can be written mathematically as:

$$\bar{X} \rightarrow N\left(\mu, \frac{\sigma^2}{n}\right) \text{as } n \rightarrow \infty$$

Alternatively,

$$z = \frac{\bar{X} - \mu}{\sigma} \sim N(0,1) \text{as } n \rightarrow \infty$$

The main significance of the central limit theorem is that it enables us to make probability statements about the sample mean when compared with the population mean. Let's distinguish two separate variables with their corresponding distributions:

1. Let X be a random variable that measures a single element from the population, then the distribution of X is the same as distribution of the population with mean (μ) and standard deviation (σ).

2. Let $\bar{X}$ be a sample mean from a population, then the distribution of $\bar{X}$ is its sampling mean with mean $(\mu_{\bar{X}})$ and standard deviation $(\sigma_{\bar{X}} = \sigma/\sqrt{n})$, where n is the sample size.

Example 8.2:

Let $\bar{X}$ be the mean of a random test of 100 customers selected from a population mean of 30 and standard deviation of 5. (a) What is the mean and standard deviation of $\bar{X}$? (b) What is the probability that the value of $\bar{X}$ falls between 29 and 31? (c) What is the probability that value of $\bar{X}$ is more than 31?

Solution

(a) $\mu_{\bar{X}} = \mu = 30; \sigma_{\bar{X}} = \dfrac{\sigma}{\sqrt{n}} = \dfrac{5}{\sqrt{100}} = 0.5$

(b) $P(29 < \bar{X} < 31) = P\left(\dfrac{29 - \mu_{\bar{X}}}{0.5} \leq z \leq \dfrac{31 - \mu_{\bar{X}}}{0.5}\right)$

$\quad = P\left(\dfrac{29 - 30}{0.5} \leq z \leq \dfrac{31 - 30}{0.5}\right) = \Phi(2) - \Phi(-2) = 0.9772 - 0.0228 = 0.9544$

(c) $P(\bar{X} > 31) = 1 - P(\bar{X} \leq 31) = 1 - P\left(\dfrac{31 - 30}{0.5}\right) = 1 - \Phi(-2) = 1 - 0.9772 = 0.0228$

R code for solutions to Example 8.2

```
Example 8.2a solution
sampl.mean<-30
mu<- sampl.mean
sample.size<-100
sigma<-5
std.dev_mean<-sigma/sqrt(sample.size)
std.dev_mean
[1] 0.5

Example 8.2b solution
sampl.mean<-mu
lower.mean<-29
z29<-(lower.mean- mu)/ std.dev_mean
p_29<-pnorm(z29, mean =0, sd = 1, lower.tail = TRUE, log.p = FALSE)
p_29
[1] 0.02275013

upper.mean<-31
z31<-(upper.mean- mu)/ std.dev_mean
p_31<-pnorm(z31, mean =0, sd = 1, lower.tail = TRUE, log.p = FALSE)
p_31
[1] 0.9772499

btw_p29_p31<- p_31- p_29
btw_p29_p31
```

[1] 0.9544997
Example 8.2c solution
p.greater31<-1-p_31
p.greater31
[1] 0.02275013

Example 8.3:

A branch manager of a microfinance bank claims that the average number of customers that deposit cash on monthly basis is 1,250 with a standard deviation of 130. Assume the distribution of cash deposits is normal, find:

(a) The probability that ≤ 1,220 customers will deposit cash in a month.

(b) The probability that given the mean for the random sample over 15 months, fewer than 1,200 customers are expected to deposit cash.

Solution

$$\text{(a) } P(X < 1220) = P\left(z \le \frac{1220 - 1250}{130}\right) = P(z \le -0.231) = 0.4090$$

$$\text{(b) } P(\bar{X} < 1200) = P\left(z \le \frac{1200 - 1250}{\frac{130}{\sqrt{15}}}\right) = P(z \le -1.49) = 0.0681$$

R code for solutions to Example 8.3

```
Example 8.3a solution
mean<-1250
std<-130
x<-1220
z<-(x- mean)/ std
p_1220<-pnorm(z, mean =0, sd = 1, lower.tail = TRUE, log.p = FALSE)
p_1220
[1] 0.408747

Example 8.3b solution
mean<-1250
std<-130
n<-15
std.err<-std/sqrt(n)
x.bar<-1200
z.xbar<-(x.bar- mean)/ std.err
p.xbar_1200<-pnorm(z.xbar, mean =0, sd = 1, lower.tail = TRUE, log.p = FALSE)
p.xbar_1200
[1] 0.06816354
```

Exercises for Chapter 8

1. (a) Explain sampling distribution.
   (b) Differentiate between probability sampling and non-probability sampling.
   (c) What are the merits and faults of probability sampling and non-probability sampling?
2. (a) State and describe the types of probability sampling techniques.
   (b) State the advantages and disadvantages of the sampling techniques mentioned in (a).
3. List non-probability sampling techniques and state the advantages and disadvantages of the techniques.
4. Using an example, explain the concept of sampling distribution of means.
5. (a) Explain the central limit theory and its importance.
   (b) Assume that the length of time of calls is normal, with an average of 60 seconds and standard deviation of 10 seconds. Find the probability that the average time obtained from a sample of 35 calls is 55 seconds.
6. Suppose a normally distributed population has a mean and standard deviation of 75 and 12, respectively.
   (a) What is the probability that a random element X selected from the population falls between 72 and 75?
   (b) Calculate the mean and standard deviation of $\bar{X}$ for a random sample of size 30.
   (c) Calculate the probability that the mean of a selected sample size of 30 from the population is between 72 and 75.

9 Confidence Intervals for Single Population Mean and Proportion

When a statistic is computed from sample data to estimate a population parameter, we have to reduce our risk of estimation by constructing a range in which the exact value of the population parameter lies. This is due to the variability in the sample collected. This process gives us the assurance within a certain percentage that the estimate is capable of predicting the real value of the population of interest. This concept describes the *confidence interval* or *confidence limit*. In this chapter, we focus on constructing the confidence interval for both mean and proportion.

9.1 Point Estimates and Interval Estimates

When a sample is drawn from a population, a single attribute obtained or quality computed from the measurement is called a *statistic*. When this statistic closely describes the characteristics obtained from the population, the estimate or a single value from the sampled data is referred to as a *point estimate*. For example, a human resource manager may want to assess the productivity of workers in a firm. The number of cases closed per month can be used as a metric to measure productivity. It is discovered that the average number of cases closed per month was 22—thus the value 22 is the point estimate.

On the other hand, in order to know how well a sample statistic accurately describes a population, we can compute a range of values within which we are confident enough to say the true value of population parameter lies. The *interval estimate* offers a measure of exactness beyond that of the point estimate by giving an interval that contains plausible values. In most cases, we use a 95% confident limit for the estimation—this implies that we can say that our estimation about a population plausibly falls within the specified range of values in 95 out of 100 cases. For instance, a researcher may like to know the number of cars traveling a road on a daily basis. A traffic study over several days concludes that the average number of cars traveling a road daily is between 180 and 200. This scenario can be represented as the normal curve shown in Figure 9.1.

https://doi.org/10.1515/9781547401475-009

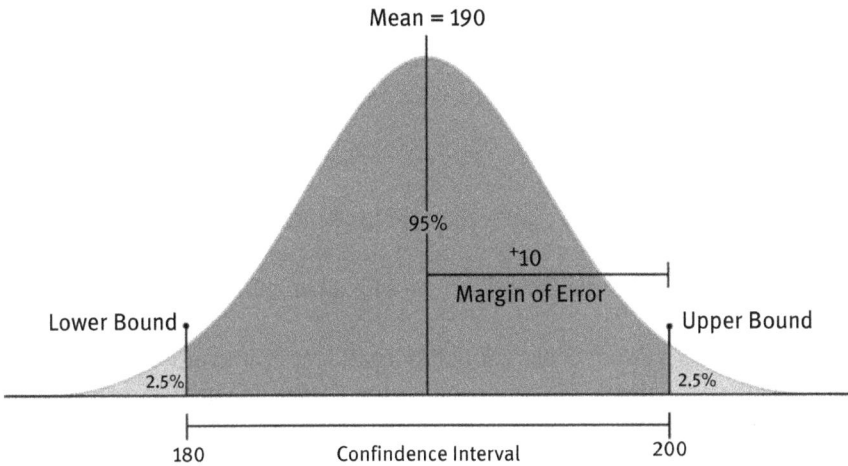

Figure 9.1: Confidence interval of the average number of cars traveling a road daily.

9.2 Confidence Intervals for Mean

If a sample is drawn many times, there is a certain percentage of assurance that a range of values computed from the sample will probably contain an unknown population parameter—this percentage is the *confidence interval*. A confidence interval is the probability that a value will fall between the upper and lower bound of a probability distribution. Suppose an analyst forecasted that the return on the Dow Jones Industrial Average (DJIA) in the first quarter of the year will fall within -0.5% and +2.5% at a 95 percent confidence interval. This means that the analyst is 95% sure that the returns on the DJIA in the first quarter of the year will lie between -0.5% and +2.5%. Confidence intervals (CIs) can be used to bound numerous statistics such as the mean, proportion, and regression coefficients (Chapter 11). A confidence interval is constructed as follows:

$$CI = sample\ statistic\ \pm\ critical\ value\ \times\ standard\ error\ of\ estimate$$

For example, if the sample size $n < 30$, σ is unknown, and the population is normally distributed, then we should use the Student's t-distribution. The *critical value* is a point on the distribution compared to the sample statistic to determine whether to accept or reject a null hypothesis.

However, if $n < 30$, σ is unknown, and the population is not normally distributed, then we should use non-parametric statistics (not covered in this book). Specifically, the confidence interval for a population mean is constructed as:

$$CI = \bar{x} \pm t_{n-1,q} \times \frac{s}{\sqrt{n}} \tag{9.1}$$

A $100(1-\alpha)\%$ confidence region for μ contains:

$$\bar{x} - t_{n-1,q} \times \frac{s}{\sqrt{n}} \leq \mu \leq \bar{x} + t_{n-1,q} \times \frac{s}{\sqrt{n}} \tag{9.2}$$

where α represents level of significance, n represents sample size, s is the standard deviation, $\bar{x}$ is the mean, t is the critical region from the t-distribution table, and q is the quantile (usually $q = 1 - \frac{\alpha}{2}$ for a two-tailed test and $q = 1 - \alpha$ for a one-tailed test).

Suppose we want to construct a 95% confidence interval for an unknown population mean, then a 95% probability that confidence interval will contain the true population mean could be calculated as follows:

$$P\left[\bar{x} - t_{n-1,\alpha/2} \times \frac{s}{\sqrt{n}} \leq \mu \leq \bar{x} + t_{n-1,\alpha/2} \times \frac{s}{\sqrt{n}}\right] = 0.95 \tag{9.3}$$

Example 9.1:
A random sample of 25 customers at a supermarket spent an average of $200 with a standard deviation of $20. Construct a 95% confidence interval estimating the population mean of purchases made at the supermarket?

Solution

$$100(1-\alpha)\% = 95\% \Rightarrow 100 - 100\alpha = 95 \Rightarrow \alpha = 0.05$$

$\bar{x} = 200$, $s = 20$, $n = 25$, $t_{n-1,1-\alpha/2} = t_{24,0.975} = 2.064$ (from t-table)

$$CI = 200 - \left(2.064 \times \frac{20}{\sqrt{25}}\right) \leq \mu \leq 200 + \left(2.064 \times \frac{20}{\sqrt{25}}\right)$$

$$CI = 191.74 \leq \mu \leq 208.26$$

This implies that the average purchase by the customers in the supermarket falls between $191.74 and $208.26.

R code for solutions to Example 9.1

```
sample.mean <- 200
sample.std <- 20
n <- 25
std.error <- qt(0.975,df=n-1)*sample.std/sqrt(n)
lower.limit <- sample.mean-std.error
upper.limit <- sample.mean+std.error
conf.interval<-c(lower.limit, upper.limit)
conf.interval
[1] 191.7444  208.2556
```

The average purchase by customers in the supermarket lies between \$191.74 and \$208.26.

In addition, if $n \geq 30$ and σ is known; or $n < 30$, σ is known, and the population is normally distributed, then we use:

$$\left[\bar{x} - z_{1-\alpha/2} \times \frac{\sigma}{\sqrt{n}} \leq \mu \leq \bar{x} + z_{1-\alpha/2} \times \frac{\sigma}{\sqrt{n}} \right] \tag{9.4}$$

If $n \geq 30$ and σ is unknown, the standard deviation s of the sample is used to approximate the population standard deviation σ, then we have:

$$\left[\bar{x} - z_{1-\alpha/2} \times \frac{s}{\sqrt{n}} \leq \mu \leq \bar{x} + z_{1-\alpha/2} \times \frac{s}{\sqrt{n}} \right] \tag{9.5}$$

Example 9.2:
A sales manager of a company wants to project a range to determine whether to expect a dramatic change in the sale of a particular product. A simple random sample of 50 sales records is taken from the previous days. The sales (in dollars) were recorded and some summary measures are provided: $n = 22$, $\bar{x} = 5200$, and $s = 400$. Assume that the sales distribution is approximately normal. (a) Construct a 95% confidence interval for the mean sales of the product. (b) Interpret your result in (a).

Solution
(a) $n = 50$, $\bar{x} = 5200$, $s = 400$, and $z_{0.025} = 1.96$

$$CI = 5200 - \left(1.96 \times \frac{400}{\sqrt{50}} \right) \leq \mu \leq 5200 + \left(1.96 \times \frac{400}{\sqrt{50}} \right)$$

$$CI = \$5089.13 \leq \mu \leq \$5310.87 \text{ or } \mu = [\$5089.13, \ \$5310.87]$$

(b) Sales of the product lie between \$5089.13 and \$5310.87

R code to calculate the confidence interval for Example 9.2

```
sample.mean <- 5200
sample.std <- 400
n <- 50
std.error <- qnorm(0.975)*sample.std/sqrt(n)
lower.limit <- sample.mean-std.error
upper.limit <- sample.mean+std.error
 conf.interval<-c(lower.limit, upper.limit)
 conf.interval
 [1] 5089.128  5310.872
```

This confirms the results in Example 9.2.

9.3 Confidence Intervals for Proportion

Suppose we are interested in estimating the proportion of people in a population with a certain qualification for a new pob posting. Let's consider educational and experience qualifications as a "success," and lack of the same qualifications as a "failure." Let X be the number of people with qualification, then the sample proportion is computed by $\hat{p} = \frac{x}{n}$, where n is the sample size.

The sampling distribution of $\hat{p}$ is approximately normal with $\mu_{\hat{p}} = p$ and $\sigma_{\hat{p}} = \frac{p(1-p)}{n}$.

Therefore, the confidence interval for the population proportion is calculated as:

$$CI = \hat{p} \pm z_{1-\frac{a}{2}} \times \sqrt{\frac{\hat{p}(1-\hat{p})}{n}} \tag{9.6}$$

The sampling distribution of $\hat{p}$ can be approximated by a normal distribution when $n\hat{p} \geq 5$ and $n\hat{p} \geq 5$.

Example 9.3:

In a an opinion poll taken to find out whether or not to market a new food product, a random sample of 8,500 participants were selected. Only 6,250 respondents approved of the product's taste while others did not. Construct a 95% confidence interval for the population proportion.

Solution

$$\text{Sample proportion}(\hat{p}) = \frac{6250}{8500} = 0.74$$

Since $n\hat{p} = 8500 \times 0.74 > 5$ and $n\hat{p} = 8500 \times 0.36 > 5$, we can use a normal distribution table. From the normal table (see Appendix), $z_{\left(1-\frac{0.05}{2}\right)} = 1.96$

$$CI = 0.74 \pm 1.96 \times \sqrt{\frac{0.74(0.26)}{8500}}$$

$$CI = 0.74 - 1.96 \times \sqrt{\frac{0.74(0.26)}{8500}} \leq p \leq 0.74 + 1.96 \times \sqrt{\frac{0.74(0.26)}{8500}}$$

$$CI = 0.7307 \leq p \leq 0.7493$$

Therefore, the proportion of respondents that supported the new product lies within 0.7307 and 0.7493.

R code for the computation of the Confidence Interval for Proportion in Example 9.3

```
sample.prop <- 0.74
n <- 8500
std.error <- qnorm(0.975)*sqrt(sample.prop*(1-sample.prop)/n)
lower.limit <- sample.prop-std.error
upper.limit <- sample.prop+std.error
```

```
conf.interval<-c(lower.limit, upper.limit)
conf.interval
[1] 0.7306752  0.7493248
```

The result shows that between 73% and 75% of the respondents favored intro-
duction of the new food product.

Example 9.4:
The manager of a commercial bank took a random sample of 120 customers' account numbers
and found that 15 customers did not have a bank verification number (BVN). Compute a 90%
confidence interval for the proportion of all the bank customers that are yet to complete the
BVN process.

Solution

$$\hat{p} = \frac{15}{120} = 0.125$$

$$CI = 0.125 \pm 1.645 \times \sqrt{\frac{0.125(0.875)}{120}}$$

$$CI = 0.125 - 1.645 \times \sqrt{\frac{0.125(0.875)}{120}} \le p \le 0.125 + 1.645 \times \sqrt{\frac{0.125(0.875)}{120}}$$

$$CI = 0.075 \le p \le 0.1747$$

The percentage of all bank customers that are yet to complete the BVN process is between
7.5% and 17.5%.

R code for the computation of the Confidence Interval for Proportion in Example 9.4

```
sample.prop <- 0.125
sample.size <- 120
std.error <- qnorm(0.95)*sqrt(sample.prop*(1-sample.prop)/sample.size)
lower.limit <- sample.prop-std.error
upper.limit <- sample.prop+std.error
conf.interval<-c(lower.limit, upper.limit)
conf.interval
[1] 0.07534126 0.17465874
```

9.4 Calculating the Sample Size

Suppose we have a given confidence level $(1-\alpha)$ as well as the margin of error
(e), and we are asked to calculate the minimum sample size. This is computed
using the general formula:

$$n \geq \left(\frac{z_{1-\frac{\alpha}{2}}}{e}\right)^2 s^2 \tag{9.7}$$

where s is the standard deviation of the sample.

Example 9.5:
A researcher claims that the standard deviation for the monthly utility bill of an individual household is \$50. He wants to estimate the mean utility bill in the present month using a 95% confidence level with a margin of error of 12. How large a sample of households is required?

Solution

$$n \geq \left(\frac{z_{0.975}}{12}\right)^2 (50)^2 = \left(\frac{1.96}{12}\right)^2 (50)^2 = 66.69 \approx 67 \, households$$

The minimum sample size required is 67 households.

R code to calculate sample size given a standard deviation and margin of error

```
pop.std<-50
margin.err<-12
z.normal<-qnorm(0.975)
sample.size<- (pop.std * z.normal/margin.err)**2
sample.size
[1] 66.69199
round(sample.size, digits = 0)
[1] 67
```

We can calculate the sample size for proportion under two conditions:
(i) When $\hat{p}$ is known, the sample size is:

$$n \geq \left(\frac{z_{1-\frac{\alpha}{2}}}{e}\right)^2 \hat{p}(1-\hat{p}) \tag{9.8}$$

Note that $\sqrt{\hat{p}(1-\hat{p})}$ is the same as the standard deviation of a proportion (s).
(ii) When $\hat{p}$ is unknown, the sample size is computed as:

$$n \geq \left(\frac{z_{1-\frac{\alpha}{2}}}{e}\right)^2 0.25 \tag{9.9}$$

This assumes that $\hat{p} = 0.5$.

Example 9.6:

In a survey, 60% of respondents supported paternal leave for male workers. Construct a 95% confidence interval for the population proportion of workers who supported this policy. The accuracy of your estimate must fall within 2.5% of the true population proportion. Find the minimum sample size to achieve this result.

Solution

$$\hat{p} = \frac{60}{100} = 0.6 \text{ and } \hat{q} = 0.4$$

To verify the sampling distribution of $\hat{p}$ to be approximated by the normal distribution, we have:

$$n\hat{p} = 100 \times 0.6 > 5 \text{ and } n\hat{q} = 100 \times 0.4 > 5$$

$$n \geq \left(\frac{z_{1-\frac{\alpha}{2}}}{e}\right)^2 \hat{p}(1-\hat{p})$$

$$n \geq \left(\frac{1.96}{0.025}\right)^2 (0.6)(0.4)$$

$$n \geq \left(\frac{1.96}{0.025}\right)^2 (0.6)(0.4) = 1475.17$$

The minimum sample size should be at least 1,475 respondents.

R code to calculate sample size given a proportion and margin of error

```
p.estimate<-0.6
q.estimate<-0.4
margin.err<-0.025
alpha<-0.05
qtile<-1-(alpha/2)
z.normal<-qnorm(0.975)
sample.size<- ((z.normal/margin.err)**2)* p.estimate* q.estimate
sample.size
[1] 1475.12
```

The survey will require a minimum of 1,475 respondents to achieve a margin of error of 2.5% at 95% CI.

9.5 Factors That Determine Margin of Error

The *margin of error* is defined as the product of critical value and standard error of the estimate. For instance, a large sample size with unknown population standard deviation has a margin of error to be $z_{1-\alpha/2} \times \frac{s}{\sqrt{n}}$ where $z_{1-\alpha/2}$ is the critical region from a normal table, s is the sample standard deviation, and n is the sample size. This margin of error is determined by the sample size, the standard deviation, and the confidence level $(1-\alpha)$.

1. When sample size increases, margin of error decreases and vice-versa.
2. The higher the standard deviation the greater the uncertainty, thus, margin of error increases.
3. The higher the confidence level, the greater the margin of error. For instance, at 95% confidence level, $z_{1-\alpha/2} = 1.96$ while at 99% confidence level $z_{1-\alpha/2} = 2.576$. The margin of error increases as the confidence level increases.

Exercises for Chapter 9

1. The grades of students in the Business Statistics course are normally distributed. If a random sample of 40 students are selected with mean 72 and standard deviation of 14, compute the 95% confidence interval for the population mean.

2. The research department of a telecommunications company wants to know the customers' usage (in hours) of a new service provided. Assuming that the usage of the service is normally distributed, a random sample of 3,000 customers using the new service is taken. They found that the mean usage is 7 hours with standard deviation of 1 hour 30 minutes. Construct a 99% confidence interval for the mean usage of the service.

3. During an economic recession period, the price of a pack of soda rose. Due to variability in the price of soda, retailers sold it at different prices. A random sample of 100 retailers were sampled with mean -$15.0 and standard deviation of $1.5. Calculate the 95% confidence interval.

4. In the process of manufacturing light bulbs, the probability that a bulb will be defective is 0.09. If a random sample of 200 bulbs is selected, compute a 95% confidence limit for the defective bulbs.

5. In order to predict the winner of the next presidential election in a country, a survey poll was conducted to allow the citizens to express their opinions about the candidates. If the 95% confidence interval is not greater than 0.09, what random sample size of respondents should be taken with the margin of error within 0.15 if the standard deviation is 20?

6. A steel rolling company manufactures cylindrical steel with the same length but different diameters (in mm). A random sample of 24 cylinders is checked and observed a mean of diameter of 12.5 mm and standard deviation of 3 mm. Compute the 95% confidence interval for the mean diameter of the cylinders.

10 Hypothesis Testing for Single Population Mean and Proportion

Hypothesis testing is used to discover whether there is enough statistical evidence in favor of or against a belief about a population parameter. Testing is used to infer results from sample data as they apply to the overall population. In this chapter, we will demonstrate how to use the concept of hypothesis testing to make a smart decision. We will also dwell on its applications in business.

10.1 Null and Alternative Hypotheses

A hypothesis is a statement speculating upon the result of a research study which can be used to describe a population parameter. The hypothesis indicating no association among groups or between measured attributes is called the *null hypothesis*; this is the hypothesis under investigation we are trying to disprove. It is denoted H_0. However, the hypothesis that observations represent the real effect of the study is called the *alternative hypothesis*. An alternative hypothesis is denoted H_1. For example, if the government wants to know if the unemployment rate in the country is different from the 5 percent claimed by the National Bureau of Statistics, the null hypothesis for this scenario is $H_0: \mu = 5\%$ versus the alternative hypothesis $H_1: \mu \neq 5\%$.

10.2 Type I and Type II Error

A *type I error* occurs when the null hypothesis is rejected when it is true. This error is also known as a *false positive*. The probability of rejecting a null hypothesis when it is true is denoted by α—i.e., P(rejecting H_0| H_0 is true) = α. Type I error is sometimes called a *producer's risk* or a *false alarm*. This error is usually set by the researcher—the lower the α value, the lower of chance of committing type I error. The probability value (*p*-value) is often set to be 0.05, except in biomedical research where the *p*-value is set to be 0.01 because it deals with human life. The probability of committing a type I error is known as the test's *level of significance*.

A *type II error* occurs when the alternative hypothesis (failing to reject the null hypothesis) is accepted as true, even though it is false. The probability of accepting an alternative hypothesis when it is actually false is denoted by β—i.e., P(accepting H_1| H_1 is false) = β. Type II error is also known as *consumer's*

https://doi.org/10.1515/9781547401475-010

risk or *misdetection*. The error is not predetermined by a researcher; rather, it is derived from the estimation of the distribution based on an alternative hypothesis and is usually unknown. The value of $1 - \beta$ is known as *power of a test*.

The power of a test gives the probability of rejecting the null hypothesis when the null hypothesis is false. To the contrary, the level of significance gives the probability of rejecting the null hypothesis when the null hypothesis is true. The values of both α and β are dependent; as one increases the other decreases. However, an increase in the sample size, n, causes both to decrease due to reduction in sampling error. Table 10.1 shows the types of errors and their respective probabilities.

Table 10.1: Statistical errors.

	H_0 rejected	Fail to reject H_0
H_0 false	Correct decision $P = 1 - \beta$ (power of a test)	Type II error $P = \beta$
H_0 true	Type I error $P = \alpha$ (level of significance)	Correct decision $P = 1 - \alpha$

10.3 Acceptance and Rejection Regions

In a hypothesis testing procedure, the sampling region is divided in two: the acceptance and rejection (critical) regions. The *acceptance region* contains a set of values for which a test statistic falls within the specified range. If the test statistic falls within the acceptance region, then the null hypothesis is accepted. On the other hand, the *rejection region* is the area of the curve where the test statistic falls outside the specified range. Thus, if the sample statistic falls within the rejection region, then the alternative hypothesis is accepted. Figure 10.1 shows the acceptance and critical region for a two-tail normal test.

10.4 Hypothesis Testing Procedures

In carrying out test of hypothesis, the following are various steps to take:

Step 1. *State the hypothesis*
A hypothesis is stated based on the argument of interest. The hypothesis can be either a one-tail test or a two-tail test. A one-tail test is a test of a hypothesis

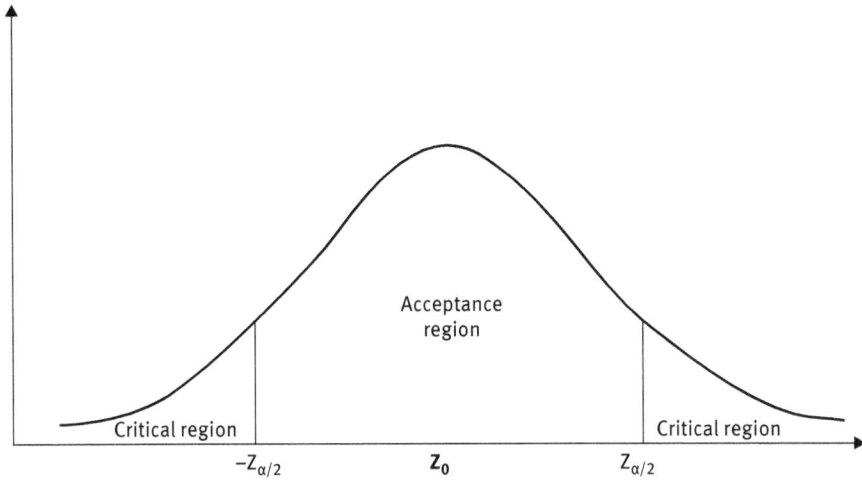

Figure 10.1: Acceptance and rejection regions.

where the region of rejection is on one side of the sampling distribution. Suppose that the null hypothesis stated that a population mean is greater than zero and the alternative hypothesis stated that a population mean is less than or equal to zero (i.e., H_0: $\mu < 0$ against H_1: $\mu \geq 0$). This implies that the rejection region would consist of a range of values located on the left side of sampling distribution—a set of values less than or equal to zero.

Alternatively, a two-tail test is a test hypothesis where the rejection region is on both sides of the sampling distribution. Assume that the null hypothesis stated that H_0: $\mu = 0$ against the alternative hypothesis H_1: $\mu \neq 0$. The non-directional sign would take the values from both sides of the sampling distribution; thus, the set of values on the right side of zero and on the left size of zero are the rejection regions.

Hypothesis testing can be of the form:

H_0: $\mu = 0$ vs. H_1: $\mu \neq 0$ (two-tail test)

H_0: $\mu = 0$ vs. H_1: $\mu < 0$ (one-tail test)

H_0: $\mu = 0$ vs. H_1: $\mu > 0$ (one-tail test)

H_0: $\mu > 0$ vs. H_1: $\mu \leq 0$ (one-tail test)

H_0: $\mu \geq 0$ vs. H_1: $\mu < 0$ (one-tail test)

H_0: $\mu < 0$ vs. H_1: $\mu \geq 0$ (one-tail test)

H_0: $\mu < 0$ vs. H_1: $\mu \geq 0$ (one-tail test)

Step 2. *Specify the level of significance*
Prior to an experiment, a researcher would choose the level of significance. In most cases, the level of signifiance is set to 10%, 5%, and 1% depending on the stringency of the research investigation. Suppose the level of significance is 5%; this indicates that there is likelihood that out of 100 cases, only 5 cases might result to rejecting the null hypothesis.

Step 3. *Compute the value of the test statistic*
The appropriate test statistic is compared with the critical value. The most commonly used test statistics are Student's t-test, normal (z-test), chi-square, and F-test. The type of statistic used depends on the properties of the data. The typical example for a test statistic is:

$$t = \frac{\sqrt{n}(\bar{x} - \mu)}{s} \sim t_{n-1, \alpha/2}$$

or

$$z = \frac{\sqrt{n}(\bar{x} - \mu)}{s} \sim z_{, \alpha/2}$$

Step 4. *Determine the critical value and decision rule*
The *critical region* is the cut-off point area of the rejection region. As long as the absolute value of the test statistic is greater than the critical value, the null hypothesis is rejected. The decision rule states that we should reject the null hypothesis if the absolute value of test statistic is higher than the critical value.

Step 5. *Draw a conclusion*
After the decision is made whether to accept or reject the null hypothesis, the next process is to draw a conclusion about the parameter under investigation.

Example 10.1:
The operations manager of a pharmaceutical company claimed that the mean dosage of a drug produced by the company is 100 milligrams. If a random sample of 60 units is chosen with mean of 98 milligrams and standard deviation of 14 milligrams, test the hypothesis to justify the operations manager's claim, use $\alpha = 0.05$.

Solution
(a) *Hypothesis*: H_0: $\mu = 100$ versus H_1: $\mu \neq 100$
(b) $\alpha = 0.05$
(c) *Test statistic*: $z = \dfrac{\sqrt{60}(98 - 100)}{14} = -1.1066$ (we use test statistic z since the sample size is greater than 30)
(d) $z_{(0.975)} = 1.96$ (from the normal distribution table, see Appendix)

The stated hypothesis is a two-tailed test, therefore we use $z_{\left(1-\frac{\alpha}{2}\right)}$. However, if the stated hypothesis is a one-tailed test and sample size is is equal to or greater than 30, then we will use $z_{(1-\alpha)}$

Decision rule: Reject the null hypothesis if $|-1.1066| > 1.96$. Since the test statistic is not greater than critical value, we do not reject H_0.

(e) *Conclusion*: The data supports the claim of the operations manager that the mean dosage of the drug produced is 100 milligrams.

The scripts below show how the question in Example 10.1 can be solved in R. The z-statistic and critical value are computed in R. This serves as a basis of comparison.

```
# state the given parameters
xbar = 98 # sample mean
mu0 = 100 # hypothesized value
sigma = 14 # sample standard deviation
n = 60 # sample size
# use z-test since sample size is greater than 30
z = (xbar-mu0)/(sigma/sqrt(n))
z # test statistic
[1] -1.106567
The z-statistic is -1.1066
# compute the critical value at 0.05 significance level
alpha = 0.05
z_alpha = qnorm(1-alpha/2)
z_alpha        # critical value
[1] 1.959964
```

The absolute value of computed z-statistic is 1.1066, which is less than the critical value of 1.96. Therefore, we do not reject the null hypothesis and we conclude that the operations manager is right in his claim with the given data.

Example 10.2:

A stockbroker claimed that weekly average return on a stock is normal with an average of return of 0.5%. He took the 20 previous weeks returns and found that the weekly average return was 0.48% with standard deviation 0.08. At a 5% level of significance, is his claim about the return accurate? If the level of significance is reduce to 1%, compare the result.

Solution

(a) *Hypothesis*: H_0: $\mu = 0.5\%$ vs. H_1: $\mu \neq 0.5\%$

(b) $\alpha = 0.05$

(c) *Test statistic*: $t = \dfrac{\sqrt{20}(0.48 - 0.50)}{0.08} = -1.118$ (we use the t-test because the sample size is less than 30). In this case, our sample size is 20.

(d) $t_{n-1,\left(1-\frac{\alpha}{2}\right)} = t_{19,(0.975)} = 2.093$ (critical value for two-tailed test is obtained from student's t-table)

Decision rule: Reject the null hypothesis if $|-1.118| > 2.093$, since the test statistic is not greater than critical value, we do not reject H_0.

(e) *Conclusion*: The stockbroker is correct on his claim based on the data. Therefore, the average weekly returns of the stock is 0.5%.

At $\alpha = 0.01$, $t = -1.118$, $t_{n-1,(1-\frac{\alpha}{2})} = t_{19,(0.995)} = 2.861$ (obtained from the student's t-table)

Decision rule: Reject null hypothesis if $|-1.118| > 2.861$, since the absolute of test statistics is less than critical value, we do not reject H_0.

Conclusion: We accept the null hypothesis based on the data, and conclude that the average weekly return is 0.5%.

Comparison: The stockbroker's claim that the average weekly return is 0.5% is correct at both 1% and 5% level of significance.

Let's demonstrate Example 10.2 with R code:

```
# Given parameters
x.bar = 0.48
mu0 = 0.50
sigma = 0.08
sample.size = 20

# compute the t-statistics
t = (x.bar-mu0)/(sigma/sqrt(sample.size ))
t
[1] -1.118034

# compute the critical value at 0.05 significance level

alpha1 = 0.05
t.alpha1 = qt(1-alpha1/2, sample.size-1 )
t.alpha1
[1] 2.093024

# At 1% level of significance
alpha2 = 0.01
t.alpha2 = qt(1-alpha2/2, sample.size-1 )
t.alpha2
[1] 2.860935
```

These results are the same as the outcomes we got in Example 10.2.

Example 10.3:

Suppose a commission conducted a test for promotion periodically, and claimed that 10% of the employees pass the test while the stakeholders argued that the percentage of employees is more than that. To ascertain the validity of the claim, a random sample of 10,000 employees are selected, of which 1,250 passed the test. Test the hypothesis that less than 10% of the employees passed the test (hint: use $\alpha = 0.05$).

Solution

(a) *Hypothesis*: H_0: $P < 10\%$ vs. H_1: $P \geq 10\%$

(b) $\alpha = 0.05$

(c) $\hat{p} = 0.125$

Test statistic: $z = \dfrac{\hat{p} - p_0}{\sqrt{\dfrac{p_0(1-p_0)}{n}}}$

Here, we used z-statistic because the sample size is 10,000 which is large, i.e., it is greater than 30.

$$z = \frac{0.125 - 0.1}{\sqrt{\frac{0.1(1-0.1)}{10000}}} = 8.3$$

(d) *Critical value*: $z_{(1-\alpha)} = z_{(0.95)} = 1.64$ (Our hypothesis is one-tailed, thus $z_{(1-\alpha)}$ is used).

Decision rule: Reject the null hypothesis if $z > z_{(0.95)}$, since 8.3>1.64, then we reject null hypothesis.

(e) *Conclusion*: There is no evidence to support the commission claim that the percentage of employees that passed the test is less than 10% based on the data.

Note: There is the possibility that another dataset might justify the claim of thecommision. However, based on the information given in this particular question, no justification for the claim that the percentage of employees who passed the test is less than 10%.

The R code below describes how the z-statistic for the proportion (p) and critical value can be obtained for proper comparison.

```
# compute the z-statistic
p.hat = 0.125
p0 = 0.10
n = 10000
z = (p.hat-p0)/ sqrt (p0*(1-p0)/n)
z
[1] 8.333333

# compute the critical value at 0.05 significance level

alpha = 0.05
z.alpha = qnorm(1-alpha)
z.alpha
[1] 1.644854
```

The z-statistic and critical value under z are 8.333333 and 1.644854, respectively.

Example 10.4:

There is a hypothesis that a region closer to the home office is more productive than one more remote. To test this hypothesis, a random sample of 1,500 employees across sectors are selected, only 1,245 employees supported the hypothesis. Use the information provided to test whether a region closer to the home office is more productive than one more remote at a 5% level of significance.

Solution

(a) Let $\hat{p}$ be the proportion of employees that supported the hypothesis under investigation.

It is expected that 50% will support the hypothesis and the remaining will be against the hypothesis.

Hypothesis: H_0: $p > 0.5$ vs. H_1: $p \leq 0.5$

(b) $\alpha = 0.05$

(c) $\hat{p} = 0.83$ and $p_0 = 0.5$

Test statistic: $z = \dfrac{\hat{p} - p_0}{\sqrt{\dfrac{p_0(1-p_0)}{n}}}$

The z-statistic is used because the sample size is greater than 30.

$$z = \frac{0.83 - 0.50}{\sqrt{\frac{0.5(1-0.5)}{1500}}} = 25.56$$

(d) *Critical value:* $z_{(1-\alpha)} = z_{(0.95)} = 1.64$ (from z-table with one-tailed test)

Decision rule: Reject the null hypothesis if $z > z_{(0.95)}$, since 25.56>1.64, thus we reject the null hypothesis.

(e) *Conclusion*: The data supports the claim that a region closer to the home office is more productive than one more remote.

R code for computing z-test for a proportion and its critical value

```
# compute the z-test for the proportion
p.cap = 0.83
p0 = 0.5
n = 1500
z = (p.cap-p0)/ sqrt (p0*(1-p0)/n)

z
[1] 25.56169

# compute the critical value at 0.05 significance level

alpha = 0.05
z.alpha = qnorm(1-alpha)
z.alpha
[1] 1.644854
```

Since both the z-test for the proportion and critical value are the same as in Example 10.4, we arrive at the same conclusion.

Exercises for Chapter 10

1. (a) Define the following:
   i.   Null hypothesis and alternative hypothesis
   ii. Type I and type II error
   (b) A company program for employees to lose weight over a three-month pe-riod, assuming that the average employee can lose 3 to 5 pounds. A ran-dom sample of 20 employees are selected and find that the mean of the weight lost is 3.95 pounds and the standard deviation is 0.85 pounds. Test the hypothesis that the weight lost by employees is greater than or

Year	1988	1989	1990	1991	1992	1993	1994	1995	1996	1997
Decrease in sales (%)	27.41	9.31	20.36	26.43	22.75	14.48	10.21	12.42	20.83	18.53

Year	1998	1999	2000	2001	2002	2003	2004	2005	2006	2007
Decrease in sales (%)	19.98	25.22	12.08	26.75	35.43	19.44	19.62	24.06	25.04	10.61

Year	2008	2009	2010	2011	2012	2013	2014	2015	2016	2017
Decrease in sales (%)	31.32	24.13	16.38	20.08	21.59	13.98	19.80	23.97	29.64	21.29

equal to 4 pounds, assuming the weight distribution is normal (use $\alpha=0.01$).

2. The managing director of a brewing company notices that sales of their products decline less than 20% during -winter months and summons the sales manager to investigate this claim. The sales manager collates the sales of products during the winter for the past 30 years. The table below shows the distribution of percentage decrease in sales during winter periods. What would you say about the managing director's claim?

3. A quality assurance manager argues that the average lifespan of light bulbs is 520 hours. To ascertain this claim a random sample of 50 bulbs is taken and the lifespan readings (in hours) are as follows:
427.82, 425.76, 395.28, 444.67, 437.26, 442.67, 424.36, 416.63, 431.49, 401.58, 407.57, 461.93, 436.58, 423.20, 429.79, 447.74, 430.27, 434.41, 414.26, 435.79, 427.24, 401.04, 433.63, 404.31, 400.14, 437.70, 437.36, 424.59, 410.77, 448.75, 421.52, 416.88, 427.79, 425.87, 412.31, 423.88, 397.12, 430.68, 418.87, 411.51 418.85, 405.59, 416.06, 388.01, 439.83, 419.70, 443.24, 422.75, 419.85, 420.28
Test whether the population mean is 520 hours at a significance level of 5%.

4. The chairman of National Union Transport & Road Workers (N.U.T.R.W.) wants to increase their fares and tells the management of the University of Lagos that the daily income of the taxi operators within the campus is less than N2,500 per day. The university formed a committee to confirm his

affirmation and took a random sample of 15 taxi operators who had average income of N2,580 with standard deviation of N200. Assuming that the income is normal distributed, justify the claim of the chairman at a level of significance of 5%.

5. a) Explain the acceptance and rejection region using examples.
    b) Discuss the procedure involved in the testing of hypotheses.
    c) The manager of a grocery store insists that a customer spends an average of $14 or more for food on a daily basis. A random sample of 18 customers are selected from the sales records and it is found that the average sale is $13.80 with variance of -$2.25. Test whether the manager's claim agrees with the sample collected. Assume that the sales of groceries is normal and use 5% significance level.

6. A beverage production plant produces 35 centiliters(cl) of drinks in bottles. The new manager suspects that the volume of the drinks do not follow the specification ascribed on the bottle. After taking a sample of a crate of the products, the following measurements were obtained:
    34.73, 36.02, 35.54, 34.72, 35.49, 35.15, 35.99, 35.35, 35.81, 34.98, 34.23, 35.45
    34.99, 34.75, 35.57, 34.40, 35.09, 34.20, 34.42, 36.09, 34.34, 35.81, 34.89
    Test the H_0: $\mu = 35$ cl against H_1: $\mu \neq 35$ cl (use $\alpha = 0.05$)

7. The dean of a polytechnic institute said that no fewer than 52% of the students that graduated from the institution fully gained empoyment immediately after leaving the school. A random sample of 1000 graduates of the institution are taken across different years to examine if they actually gained employment immediately after leaving the school. The table below shows the number of the graduates from the institute and employment status immediately after graduation.

Employment status	Employed	Not employed
Number of graduated students	542	458

Test whether the dean is correct in his statement and test H_0: $p \geq 0.5$ against H_1: $p < 0.5$, use $\alpha = 0.05$.

11 Regression Analysis and Correlation Analysis

This chapter will focus on how two or more variables are interrelated. By the end of this chapter, readers will understand the concept of *regression analysis* (the study of relationships between a set of variables) and *correlation analysis* (the analysis of closeness between variables). We discuss what to look for in the output of regression analysis and how the output can be interpreted. We will demonstrate how to use regression analysis for forecasting.

11.1 Construction of Line Fit Plots

In showing the relationship between two variables, we can draw a line across the variables after plotting the scatter plot and ensure the line passes through as many points as possible. The straight line that gives the best approximation in a given set of data is referred to as the *line of best fit*. The *least squares method* is the most accurate for finding the line of best fit of a given dataset.

For example, Table 11.1 shows the sales revenue ($'million) and amount spent on advertisement ($'million) of a production company. The relationship between sales revenue and advertising expenses is shown in Figure 11.1. The equation on the line of best fit is sales = advert +111. This implies that when there are no expenses incurred on advertisement, the sales revenue will be $111 million. As advertising expenses increase, the sales revenue also increases.

11.2 Types of Regression Analysis

There are many types of *regression analysis* and they are based on different assumptions. This book will focus on the first two types in this list:
- Simple linear regression
- Multiple regression
- Ridge regression
- Quantile regression
- Bayesian regression

https://doi.org/10.1515/9781547401475-011

Table 11.1: Revenue and expenses of a production company.

Sales revenue ($'million)	115	118	120	125	126	128	131	132
Advertising expenses ($'million)	4	7	9	14	15	17	20	21

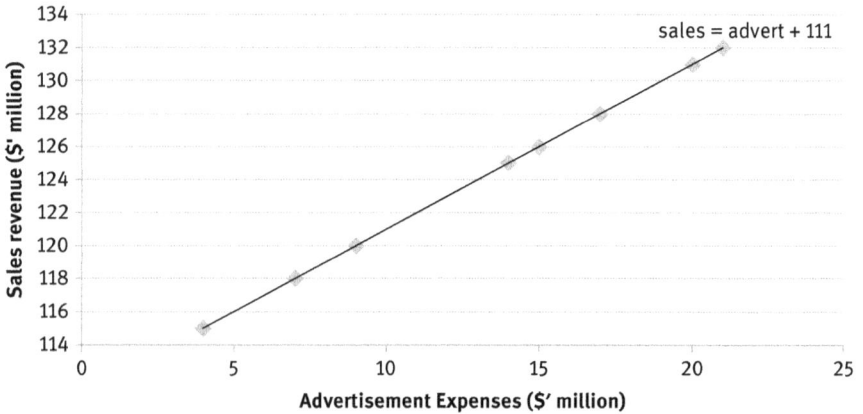

Figure 11.1: Chart of sales revenue and advertising expenses.

11.2.1 Uses of Regression Analysis

Regression analysis can be used for the following:
- Causal analysis—to establish a relationship between two or more variables, with independent variables considered as the cause of a dependent variable.
- Forecasting an effect—to predict a response variable fully knowing the independent variables.
- Forecasting a trend—to predict trend in a dataset.

11.3 Simple Linear Regression

Simple linear regression is a statistical technique to show the relationship between one dependent variable and one independent variable. The dependent variable is denoted Y while the independent variable is denoted X. The variables X and Y are linearly related. Simple linear regresson can be used for: (a) description of the linear dependence of one variable to another variable; (b) prediction of one variable from the values of another variable; or (c) correction for the linear dependence of one variable on another variable.

The simple linear regression model is of the form:

$$Y_i = \beta_0 + \beta_1 X_i + \varepsilon_i \tag{11.1}$$

where β_0 and β_1 are the intercept and regression coefficient of X and ε is the error term.

The solution to the regression coefficients in (11.1) can be derived using the *Least Square Method*:

$$e_i = Y_i - \beta_0 - \beta_1 X_i \tag{11.2}$$

To find a minimum sum of squares of residuals (the bits left over off the least squares line) you set the sum below equal to zero:

$$\sum_{i=1}^{n} (e_i)^2 = \sum_{i=1}^{n} (Y_i - \beta_0 - \beta_1 X_i)^2 = 0 \tag{11.3}$$

Taking the partial derivative of (11.3) with respect to β_1:

$$\frac{\delta}{\delta \beta_0} \sum_{i=1}^{n} (Y_i - \beta_0 - \beta_1 X_i)^2 = -2 \left(n\beta_0 + \beta_1 \sum_{i=1}^{n} X_i - \sum_{i=1}^{n} Y_i \right) = 0 \tag{11.4}$$

Divide (11.4) by 2, and solve for β_0.
Then,

$$\beta_0 = \bar{Y} - \beta_1 \bar{X} \tag{11.5}$$

Now,

$$\frac{\delta}{\delta \beta_1} \sum_{i=1}^{n} (Y_i - \beta_0 - \beta_1 X_i)^2 = -2 \sum_{i=1}^{n} (X_i Y_i - \beta_0 X_i - \beta_1 X_i^2) = 0 \tag{11.6}$$

Substitute for the value of β_0 in (11.6).

$$\sum_{i=1}^{n} (X_i Y_i - (\bar{Y} - \beta_1 \bar{X}) X_i - \beta_1 X_i^2) = 0$$

$$\sum_{i=1}^{n} (X_i Y_i - X_i \bar{Y} - \beta_1 X_i \bar{X} - \beta_1 X_i^2) = 0$$

$$\sum_{i=1}^{n} (X_i Y_i - X_i \bar{Y}) - \beta_1 \sum_{i=1}^{n} (X_i^2 - X_i \bar{X}) = 0$$

Therefore,

$$\beta_1 = \frac{\sum_{i=1}^{n}(X_iY_i - X_i\bar{Y})}{\sum_{i=1}^{n}(X_i^2 - X_i\bar{X})} = \frac{\sum_{i=1}^{n}(X_iY_i) - n\bar{X}\bar{Y}}{\sum_{i=1}^{n}(X_i^2) - n\bar{X}^2} = \frac{cov(X,Y)}{var(X)} \tag{11.7}$$

where $cov(X,Y)$ is the covariance of X and Y—a measure of how X varies with Y.

11.3.1 Assumptions of Simple Linear Regression

- There is a linear relationship between variable Y and variable X
- Variable X is deterministic or non-stochastic
- The error terms are statistically independent
- There is no correlation between X and ε_i, that is, $Cov(X, \varepsilon_i) = 0$
- The error terms are distributed normally with mean zero and constant variance, i.e., $\varepsilon\tilde{N}(0, \ \sigma^2)$
- There is no correlation between the error terms, i.e., no serial auto-correlation in the data
- The number of sample observations must be greater than the number of parameters to be estimated
- For each value of X, the distribution of residuals has equal variance, i.e., homoscedasticity

---

Example 11.1:
In a business statistics class, the weight and height of 30 students were measured and are shown in Table 11.2.

Table 11.2: The height and weight measurements of 30 students in a business statistics class.

Student	1	2	3	4	5	6	7	8	9
Height (m)	1.43	1.10	1.24	1.36	2.26	1.25	1.74	1.55	1.51
Weight (kg)	92.18	77.76	65.44	114.19	82.81	106.66	94.44	75.32	67.35

Student	10	11	12	13	14	15	16	17	18
Height (m)	1.82	1.57	1.59	2.19	1.54	2.06	1.86	1.76	1.51
Weight (kg)	101.55	76.37	91.66	75.85	88.82	83.02	74.66	97.57	104.56

Student	19	20	21	22	23	24	25	26	27
Height (m)	2.39	1.83	2.02	1.99	1.40	1.54	1.60	1.88	1.52
Weight (kg)	113.36	64.71	103.79	70.02	78.35	80.70	90.54	91.55	82.57

Student	28	29	30
Height (m)	1.41	1.38	1.18
Weight (kg)	82.49	87.98	67.54

(i) Find the regression of the weight versus the height of the students.
(ii) Use your answer in (i) to estimate the value of a student's weight when the height is 1.80.

Solution
From the data above, we obtained the following results:

$$\sum XY = 4282.115, \sum Y = 2583.81, \sum X = 49.48, n = 30, \bar{X} = 1.6493,$$

$$\bar{Y} = 86.127, \sum X^2 = 84.699$$

Substitute for these values in equation (11.7),

$$\beta_1 = \frac{\sum_{i=1}^{n}(X_iY_i) - n\bar{X}\bar{Y}}{\sum_{i=1}^{n}(X_i^2) - n\bar{X}^2} = \frac{4282.115 - 30(1.6493)(86.13)}{84.699 - 30(1.6493)^2} = 6.6503$$

The slope of the regression is 6.6503.
Then substitute for $\bar{Y}, \bar{X},$ and β_1 in equation 11.5, to get the constant term:

$$\beta_0 = \bar{Y} - \beta_1\bar{X} = 86.13 - (6.6503)(1.6493) = 75.1587$$

(i) The regression model is Weight = 75.16 + 6.65*Height
(ii) Weight = 75.16 + 6.65*1.80 = 87.13

To estimate the regression estimates in R
Assign values for height and weight in R as follows:

```
height<-c(1.43, 1.10, 1.24, 1.36, 2.26, 1.25, 1.74, 1.55, 1.51, 1.82, 1.57, 1.59,
2.19, 1.54, 2.06, 1.86, 1.76, 1.51, 2.39, 1.83, 2.02, 1.99, 1.40, 1.54, 1.60,
1.88, 1.52, 1.41, 1.38, 1.18)
weight<-c(92.18, 77.76, 65.44, 114.19, 82.81, 106.66, 94.44, 75.32, 67.35,
101.55, 76.37, 91.66, 75.85, 88.82, 83.02, 74.66, 97.57, 104.56, 113.36, 64.71,
103.79, 70.02, 78.35, 80.70, 90.54, 91.55, 82.57, 82.49, 87.98, 67.54)
```

Fit the model by regressing weight versus height of the students and the results
will follow.

```
model.fit<-lm (weight~height)
summary(model.fit)

Call:
lm(formula = weight ~ height)

Residuals:
Min 1Q Median 3Q Max
-22.619 -9.917 -2.371 7.660 29.987

Coefficients:
Estimate Std. Error t value Pr(>|t|)
(Intercept) 75.159 13.433 5.595 5.47e-06 ***
height 6.650 7.995 0.832 0.413
```

Signif. codes: 0 '***' 0.001 '**' 0.01 '*' 0.05 '.' 0.1 ' ' 1
Residual standard error: 14.05 on 28 degrees of freedom
Multiple R-squared: 0.02412, Adjusted R-squared: -0.01074
F-statistic:0.6919 on 1 and 28 DF, p-value: 0.4125

In the results above, the statistics for residuals, the coefficient estimates (with standard errors and the associated p-values), and all other statistics (Multiple R-squared, Adjusted R-squared, F-statistics, etc.) are shown in the output.

Example 11.2:

Table 11.3 shows the log of gross domestic products (LNGDP) and the log of government spending (LNGEXP) in Nigeria between 1981–2015. Regress the log of gross domestic product (LNGDP) on the log of government spending (LNGEXP) and interpret your result.

Table 11.3: The data for Example 11.2.

Year	1981	1982	1983	1984	1985	1986	1987	1988	1989	1990
LNGDP (Y)	25.545	25.535	25.483	25.463	25.543	25.451	25.337	25.410	25.473	25.593
LNGEXP (X)	21.082	21.105	21.128	21.150	21.171	21.192	21.213	21.233	21.253	21.273

Year	1991	1992	1993	1994	1995	1996	1997	1998	1999	2000
LNGDP (Y)	25.587	25.591	25.612	25.621	25.618	25.666	25.694	25.721	25.725	25.777
LNGEXP (X)	21.283	21.312	21.340	21.354	21.354	21.382	21.399	21.416	21.433	21.449

Year	2001	2002	2003	2004	2005	2006	2007	2008	2009	2010
LNGDP (Y)	25.820	25.858	25.956	26.247	26.281	26.360	26.426	26.486	26.554	26.629
LNGEXP (X)	21.320	21.377	21.103	22.998	23.098	23.404	23.854	24.069	24.076	24.188

Year	2011	2012	2013	2014	2015
LNGDP (Y)	26.677	26.719	26.771	26.832	26.859
LNGEXP (X)	24.233	24.213	24.105	24.032	24.028

Source: *World Development Indicator.*

Solution

From Table 11.3, we obtain the following because they are needed in the computation of the regression coefficients (β_0 and β_1):

$$\sum^X Y = 20140.45, \sum Y = 907.92, \sum X = 775.62, n = 35, \bar{X} = 22.16, \bar{Y} = 25.94, \sum X^2 = 17243.21$$

Substitute for these values to compute the regression coefficients (β_0 and β_1):

$$\beta_1 = \frac{\sum_{i=1}^{n}(X_i Y_i) - n\bar{X}\bar{Y}}{\sum_{i=1}^{n}(X_i^2) - n\bar{X}^2} = \frac{20140.45 - 35(22.16)(25.94)}{17243.21 - 35(22.16)^2} = 0.382$$

$$\beta_0 = \bar{Y} - \beta_1 \bar{X} = 25.94 - (0.382)(22.16) = 17.475$$

Thus, the regression model is LNGDP $= 17.475 + 0.382*$LNGEXP

Interpretation: This implies that a 1 percent increase in government spending would lead to a 0.4 percent increase in gross domestic product.

R code for regressing the log of gross domestic product (LNGDP) on the log of government spending (LNGEXP)

Assign the values for the log of gross domestic product (LNGDP) and the log of government spending (LNGEXP):

LNGDP<-c(25.545, 25.535, 25.483, 25.463, 25.543, 25.451, 25.337, 25.410, 25.473, 25.593, 25.587, 25.591, 25.612, 25.621, 25.618, 25.666, 25.694, 25.721, 25.725, 25.777, 25.820, 25.858, 25.956, 26.247, 26.281, 26.360, 26.426, 26.486, 26.554, 26.629, 26.677, 26.719, 26.771, 26.832, 26.859)

LNGEXP<-c(21.082, 21.105, 21.128, 21.150, 21.171, 21.192, 21.213, 21.233, 21.253, 21.273, 21.283, 21.312, 21.340, 21.354, 21.354, 21.382, 21.399, 21.416, 21.433, 21.449, 21.320, 21.377, 21.103, 22.998, 23.098, 23.404, 23.854, 24.069, 24.076, 24.188, 24.233, 24.213, 24.105, 24.032, 24.028)

```
# store LNGDPand LNGEXP as a data frame into data
data<-data.frame (LNGDP, LNGEXP)
model<-lm(LNGDP~ LNGEXP, data)
summary(model)

Call:
lm(formula = LNGDP ~ LNGEXP)

Residuals:
Min 1Q Median 3Q Max
-0.25038 -0.06998 -0.01893 0.04646 0.40962

Coefficients:
Estimate Std. Error t value Pr(>|t|)
(Intercept)  17.68058 0.39751 44.48 <2e-16 ***
LNGEXP 0.37273 0.01791 20.81 <2e-16 ***
---
Signif. codes: 0 '***' 0.001 '**' 0.01 '*' 0.05 '.' 0.1 ' ' 1

Residual standard error: 0.1328 on 33 degrees of freedom
Multiple R-squared: 0.9292, Adjusted R-squared: 0.9271
F-statistic: 433.2 on 1 and 33 DF, p-value: < 2.2e-16
```

From the output above, the regression coefficients are 17.68 and 0.37 for the intercept and slope, respectively. The constant (intercept) term and the coefficient of LNGEXP are significant as the Pr(>|t|) = 2e-16 each, this value is less

than a 5% level of significance. The degree of freedom is the difference between the number of observations (n) and the number of the estimated parameters (β_0 and β_1). That is, $35-2 = 33$ degrees of freedom. The coefficient of
determination, *Adjusted R-squared* indicates that 92% of the of the sample variability in LNGDP can be explained by the model. The F-statistic shows that all
the parameters are jointly significant as *p-value: < 2.2e-16* corresponding to F
less than 5%.

11.4 Multiple Linear Regression

In simple linear regression, we considered two variables where one is the response and the other is the explanatory variable. *Multiple linear regression* is an
extension of simple linear regression where we have two or more independent
variables that account for the variation in a dependent variable. Each of the explanatory variables X_i is associated with a value of the response variable Y. The
multiple linear regression model is of the form:

$$Y_i = \beta_0 + \beta_1 X_{i1} + \beta_2 X_{i2} + \beta_3 X_{i3} + \ldots + \beta_k X_{ik} + \varepsilon_{ik}, \text{ for } i = 1, 2, \ldots, n \qquad (11.8)$$

where β_0 is a constant term, β_1, β_2, $\ldots$, β_k are regression coefficients, ε_{ik} is the
error term and $\varepsilon_{ik} \sim N(0, \sigma^2)$.

The estimates of the regression coefficients (β_0, β_1, β_2, $\ldots$, β_k) are the values that minimize the sum of squared errors for the residuals.

For two independents variables,

$$Y_i = \beta_0 + \beta_1 X_{i1} + \beta_2 X_{i2} + \varepsilon_{ik} \qquad (11.9)$$

The regression coefficients can be estimated as:

$$\beta_1 = \frac{\left(\sum x_2^2\right)\left(\sum x_1 y\right) - \left(\sum x_1 x_2\right)\left(\sum x_2 y\right)}{\left(\sum X_1^2\right)\left(\sum X_2^2\right) - \left(\sum X_1 X_2\right)^2} \qquad (11.10)$$

$$\beta_2 = \frac{\left(\sum x_1^2\right)\left(\sum x_2 y\right) - \left(\sum x_1 x_2\right)\left(\sum x_1 y\right)}{\left(\sum x_1^2\right)\left(\sum x_2^2\right) - \left(\sum x_1 x_2\right)^2} \qquad (11.11)$$

$$\beta_0 = \bar{Y} - \beta_1 \bar{X}_1 - \beta_2 \bar{X}_2 \qquad (11.12)$$

where

$$\sum x_1 y = \sum X_1 Y - \frac{\left(\sum X_1\right)\left(\sum Y\right)}{N} \qquad (11.13)$$

$$\sum x_2 y = \sum X_2 Y - \frac{(\sum X_2)(\sum Y)}{N} \tag{11.14}$$

$$\sum x_1 x_2 = \sum X_1 X_2 - \frac{(\sum X_1)(\sum X_2)}{N} \tag{11.15}$$

Interpretation: β_0 is the constant term or intercept at Y. β_1 is the change in Y for each 1 unit change in X_1 while X_2 is held constant, and β_2 is the change in Y for each 1 unit change in X_2 holding X_1 constant.

Example 11.3:
Table 11.4 shows the level of education (X_1), years of experience (X_2), and the log of monthly compensation of a company (Y). The level of education $X_1 = 1$ for primary education, $X_1 = 2$ for secondary education, $X_1 = 3$ for a polytechnic graduate, $X_1 = 4$ for a university bachelor degree, and $X_1 = 5$ for a university masters degree holder. Regress the log of monthly compensation (Y) on the level of education (X_1) and the years of experience in the company (X_2).

Table 11.4: The level of education, experience, and compensation structure of a company.

Level of education (X_1)	Experience (X_2)	Log of Compensation (Y)
1	8	6.43
2	4	6.73
2	6	6.76
4	2	7.00
3	7	7.10
4	5	7.34
5	5	7.53
3	3	6.75
4	2	6.83
2	6	7.00
2	3	6.51
3	5	6.82
3	8	6.94
1	5	6.23
3	5	6.82
5	5	7.53
5	3	7.31
5	4	7.43
4	5	7.34
4	5	7.34
2	3	6.51
4	9	7.61

Table 11.4 (continued)

Level of education (X_1)	Experience (X_2)	Log of Compensation (Y)
3	2	6.71
5	4	7.43
3	4	6.79
1	7	6.40
4	5	7.34
2	3	6.51
2	9	7.17
2	5	6.75
5	6	7.55
1	2	5.92
1	5	6.23
3	3	6.75
4	4	7.31

Solution
We obtained the following values from the data above

$$\sum X_2^2 = 921, \sum X_1^2 = 387, \sum X_1 Y = 759.61, \sum X_2 Y = 1165.55,$$
$$\sum X_1 X_2 = 499, n = 35, \sum X_1 = 107, \sum X_2 = 167, \sum Y = 242.76$$

Compute the following and substitute the values for the regression coefficients obtained:

$$\sum x_1 y = 759.61 - \frac{(107)(242.76)}{35} = 17.458$$

$$\sum x_2 y = 1165.55 - \frac{(167)(242.76)}{35} = 7.238$$

$$\sum x_1 x_2 = 499 - \frac{(107)(167)}{35} = -11.543$$

$$\sum x_2^2 = 921 - \frac{(167)(167)}{35} = 124.171$$

$$\sum x_1^2 = 387 - \frac{(107)(107)}{35} = 59.886$$

$$\beta_1 = \frac{\left(\sum x_2^2\right)\left(\sum x_1 y\right) - \left(\sum x_1 x_2\right)\left(\sum x_2 y\right)}{\left(\sum x_1^2\right)\left(\sum x_2^2\right) - \left(\sum x_1 x_2\right)^2} = \frac{(124.171)(17.458) - (-11.543)(7.238)}{(59.886)(124.171) - (-11.543)^2} = 0.308$$

$$\beta_2 = \frac{(59.886)(7.238) - (-11.543)(17.458)}{(59.886)(124.171) - (-11.543)^2} = 0.084$$

$$\beta_0 = 6.94 - 0.308(3.06) - 0.084(4.77) = 5.597$$

The regression coefficients are 5.597, 0.308, and 0.084 for β_0, β_1, and β_2, respectively. The model is $Y_i = 5.577 + 0.31X_{i1} + 0.087X_{i2}$

Interpretation: This implies that a 1 unit increase in the level of education would lead to a 31 percent increase in compensation holding the years of experience constant. In addition, every 1 unit increase in years of experience would lead to a 9 percent increase in compensation holding level of education constant.

R code for the solution of the Example 11.3

Input the values of level of education (edu), year of employment (empl), and log of compensation (comp)

```
edu<-c(1, 2, 2, 4, 3, 4, 5, 3, 4, 2, 2, 3, 3, 1, 3, 5, 5, 5, 4, 4, 2, 4, 3, 5, 3,
1, 4, 2, 2, 2, 5, 1, 1, 3, 4)
```

```
empl<-c(8, 4, 6, 2, 7, 5, 5, 3, 2, 6, 3, 5, 8, 5, 5, 5, 3, 4, 5, 5, 3, 9, 2, 4, 4,
7, 5, 3, 9, 5, 6, 2, 5, 3, 4)
```

```
comp<-c(6.43, 6.73, 6.76, 7, 7.10, 7.34, 7.53, 6.75, 6.83, 7, 6.51, 6.82, 6.94,
6.23, 6.82, 7.53, 7.31, 7.43, 7.34, 7.34, 6.51, 7.61, 6.71, 7.43, 6.79, 6.40, 7.34,
6.51, 7.17, 6.75, 7.55, 5.92, 6.23, 6.75, 7.31)
```

Combine the variables (edu, empl, and comp) as a dataframe and then regress comp on edu and empl:

```
mydata<-data.frame (edu, empl, comp)
multiple.reg<-lm(comp~ edu+empl, mydata)
```

Obtain the regression coefficient and other statistics as follow:

```
summary(multiple.reg)
```

```
Call:
lm(formula = comp ~ edu + empl, data = mydata)
```

```
Residuals:
Min 1Q Median 3Q Max
-0.25870 -0.09077 -0.01283 0.07499 0.28368
```

Coefficients:	Estimate	Std. Error	t value	Pr(>|t|)	
(Intercept)	5.57723	0.07656	72.844	< 2e-16	***
edu	0.30803	0.01540	20.002	< 2e-16	***
empl	0.08717	0.01069	8.151	2.61e-09	***

Signif. codes: 0 '***' 0.001 '**' 0.01 '*' 0.05 '.' 0.1 ' ' 1

```
Residual standard error: 0.1181 on 32 degrees of freedom
Multiple R-squared: 0.9308, Adjusted R-squared: 0.9265
F-statistic: 215.3 on 2 and 32 DF, p-value: < 2.2e-16
```

The regression coefficients are the same as Example 11.3 and are interpreted in the same manner.

11.4.1 Significance Testing of Each Variable

In this subsection, we are testing whether the independent variables in the model are useful, i.e., if the independent variables can usually help us predict the dependent variable (Y). In order to determine whether variable X_1 is making a useful contribution in the model, we have to test its significance by setting the hypotheses as shown:

$Hypothesis$: H_0: $\beta_1 = 0$ vs. H_1: $\beta_1 \neq 0$

$Test\ statistic$: $t = \frac{\beta_1}{se(\beta_1)} \sim t_{1-\frac{\alpha}{2},\ n-2}$

$Decision\ rule$: Reject H_0 if $|t| \geq t_{1-\frac{\alpha}{2},\ n-2}$

Example 11.4:
From the information in Example 11.3, test the null hypothesis that β_1 is significantly different from zero, i.e., H_0: $\beta_1 = 0$ vs. H_1: $\beta_1 \neq 0$

Solution
$Hypothesis$: H_0: $\beta_1 = 0$ vs. H_1: $\beta_1 \neq 0$
$Test\ statistic$: $t = \frac{0.30803}{0.01540} = 20.002$
$Critical\ value$: $t_{0.975,\ 33} = 2.042$
$Decision\ rule$: Reject H_0 if $|t| \geq t_{0.975,\ 33}$—since $20.002 \geq 2.042$, we reject H_0
$Conclusion$: The coefficient of the level of education X_1 is significantly different from zero.
Note: This result gives the same conclusion as with the model in Example 11.3.

11.4.2 Interpretation of Regression Coefficients and Other Output

1. Regression coefficients

The magnitude of the coefficient of each independent variable gives the size of the effect that the independent variable has on the dependent variable. The sign (-/+) on the regression coefficient expresses the direction of the effect. In general, the coefficient reveals *how much* the dependent variable would change when the independent variable changes by 1 unit, keeping all other independent variables constan t.

For example, if we fit a model, $Y = 2.5 + 0.5X_1 - 0.8X_2$, it can be interpreted as follows: The Y-intercept can be intepreted as the predicted value for Y when both X_1 and X_2 are zero. Therefore, you would expect 2.5 unit value for Y when $X_1 = 0$ and $X_2 = 0$. Every 1 unit increase in Y would lead to a 0.5 unit increase in X_1 while X_2 is held constant, and every 1 unit increase in Y would lead to a 0.8 unit decrease in X_2 holding X_1 constant.

2. t-value

This is the ratio of the coefficient and the standard error of the coefficient. The rule of thumb is that the absolute value of t must be 2 or more to show the significance of the coefficient. T-value is used to determine the p-value correspondingto the Student's t-distribution.

3. P-value

This indicates the probability that the estimated coefficient is not reliable. The less the p-value, the more reliable it is under the significance level. For example, if the level of significance is 5% or (10%) a value of p less than 5% or (10%) indicates that the estimated coefficient is reliable, otherwise it is unreliable and it should be discarded from the model.

4. Multiple R-squared

This shows the fraction (percentage) of the variation in a response variable that is accounted for by independent variables in the model. It indicates how well the terms fit the data. In addition, the adjusted R-squared is used to adjust for the number of variables in a model. As you add more independent variables to a model, the R-squared continues to increase in value, even when a variable is not directly applicable in the model. However, the *adjusted* R-squared will only increase if you add a useful independent variable to the model, otherwise the value of adjusted R-squared will decrease. The R-squared ranges from zero (0) to 1 but the adjusted R-squared can dip down to a negative value.

5. F-statistics

This tests for the significance of the overall coefficients and whether the regression model provides a better fit to the data than a model with no independent variables.

6. Durbin-Watson

This is used to test for the autocorrelation assumption of the error terms. That is, to make sure that there is no correlation between the error terms $Cov(\varepsilon_i, \varepsilon_{i-1}) = 0$. *Autocorrelation* may be caused by omission of an important explanatory variable, misspecification of the model, and/or a systematic error in measurement. The consequences of autocorrelation include that the least square estimators will be inefficient and/or the estimated variances of the regression coefficients will be biased and inconsistent, thus hypothesis testing is no longer valid. Furthermore, Durbin-Watson ranges from 0 to 4. The value of 2 indicates no autocorrelation between the error terms, between 0 and 2 is a positive autocorrelation and beween 2 and 4 is a negative autocorrelation. A rule of thumb for Durbin-Watson is that for relatively normal data, the test statistic should fall within 1.5 and 2.5.

Regression output in R

After inputting the values for dependent and independent variables, and combining the series and storing as a dataframe, then the R function *lm()* is used to regress the dependent variable on the set of independent variables. The function *summary()* gives the result below as explained in the Example 11.2. We discuss the most important results in this output. The regression coefficients of the model are 17.68 and 0.37 with the corresponding *p*-values of 2e-16 each, this is very close to zero and much less than 5%. This indicates that the coefficients are reliable. Adjusted R-squared (0.93) indicates that a 93% variation in LNGDP is explained by the model. This shows that the model fits well. Also, the *p*-value of F-statistics is 2.2e-16, indicating that all the regression coefficients (intercept and slope) are jointly significant.

```
Call:
lm(formula = LNGDP ~ LNGEXP)

Residuals:
Min 1Q Median 3Q
Max
-0.25038 -0.06998 -0.01893 0.04646 0.40962

Coefficients:
Estimate Std. Error t value Pr(>|t|)
(Intercept)  17.68058 0.39751 44.48 <2e-16 ***
LNGEXP 0.37273 0.01791 20.81
<2e-16 ***
---
Signif. codes: 0 '***' 0.001 '**' 0.01 '*' 0.05 '.' 0.1 ' ' 1

Residual standard error: 0.1328 on 33 degrees of freedom
Multiple R-squared: 0.9292, Adjusted R-squared: 0.9271
F-statistic: 433.2 on 1 and 33 DF, p-value: < 2.2e-16
```

11.5 Pearson Correlation Coefficient

A simple linear regression analysis shows the relationship between two variables that are linearly related. The correlation analysis is a measure of strength or level of association between variables. The correlation is measured by the *Pearson correlation coefficient* and is denoted by "r." The statistic *r* ranges from -1 to +1. If the value of *r* is zero, it implies that there is no linear association between the variables. As the value of the correlation coefficient *r* moves close to zero (0), the linear association between the variables becomes weaker. Conversely, as the value of the correlation coefficient *r* moves far away from zero and approaches ±1, the linear association between the variables becomes stronger. When the correlation coefficient is exactly 1, it is called a *perfect positive correlation* and when the correlation coefficient is exactly -1, then it is known as a *perfect negative correlation*.

-1	0	1
Perfect negative association Negative association	No association	Perfect positive Positive association association

The significance of a relationship as determined whether the Pearson's correlation coefficient is a meaningful reflection of the linear relationship between two variables or whether the relationship occurred by chance. For a given significant value ($\alpha = 0.05$), the probability that Pearson's correlation coefficient value comes by chance is 5% or less.

Pearson's correlation coefficient *r* between variables X and Y can be defined as:

$$r_{XY} = \frac{\sum_{i=1}^{n}(X_i - \bar{X})(Y_i - \bar{Y})}{\sqrt{\sum_{i=1}^{n}(X_i - \bar{X})^2}\sqrt{\sum_{i=1}^{n}(Y_i - \bar{Y})^2}} \tag{11.16}$$

Alternatively,

$$r_{XY} = \frac{n\sum_{i=1}^{n}X_iY_i - \left(\sum_{i=1}^{n}Y_i\right)\left(\sum_{i=1}^{n}X_i\right)}{\sqrt{\left[n\sum_{i=1}^{n}X^2 - \left(\sum_{i=1}^{n}X_i\right)^2\right]\left[n\sum_{i=1}^{n}Y^2 - \left(\sum_{i=1}^{n}Y_i\right)^2\right]}} = \frac{Cov(X,Y)}{S_X.S_Y} \tag{11.17}$$

where $\bar{X}$ and $\bar{Y}$ are the means of *X* and *Y*, respectively.
   $Cov(X, Y)$ is the covariance between *X* and *Y*.
   S_X and S_Y are the standard deviations of *X* and *Y*. respectively.

11.5.1 Assumptions of Correlation Test

The following are the assumptions under the Pearson correlation test:
- There must be independent observations
- The population correlation is assumed to be zero, i.e., $\rho = 0$
- The bivariates are normally distributed in the population

11.5.2 Types of Correlation

The different types of correlation are displayed in the various charts in Figure 11.2. These include perfect positive correlation, perfect negative correlation, low degree of positive correlation, low degree negative correlation, high degree of positive correlation, high degree of negative correlation and no correlation graphs.

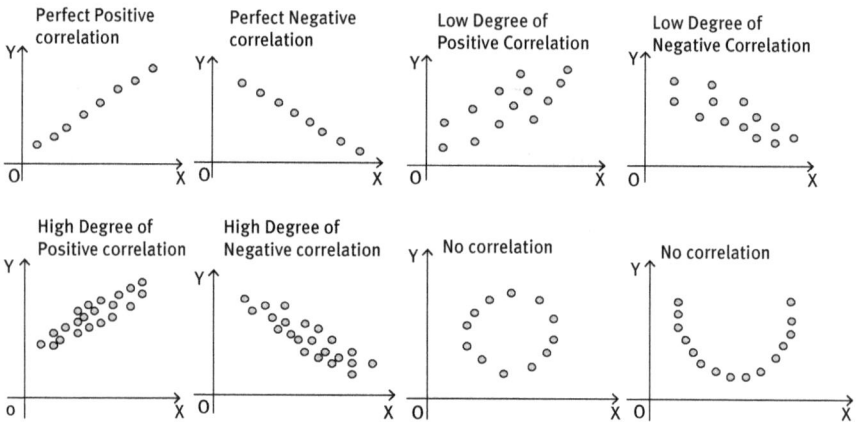

Figure 11.2: Types of correlation.
Source: https://www.slideshare.net/RamKumarshah/correlation-and-regression-56561989.

11.5.3 Coefficient of Determination

The *coefficient of determination* is defined as the square of the correlation coefficient and is denoted r^2. It measures the strength of the relationship between two variables. The coefficient of determination is a measure of the ratio of proportion variance explained by the model. That is, it measures the percentage of variation in the dependent variable that is accounted for by the variation in the independent variable. For instance, if $r = 0.9$, then the coefficient of determination $r^2 = 0.81$. This indicates that an 81% variation in the dependent variable is accounted for by variation in the independent variable. Thus, the remaining 19% variation in the dependent variable cannot be explained by the variation in the independent variable. There must be another variable to be found.

11.5.4 Test for the Significance of Correlation Coefficient (r)

This significance test is used to test the null hypothesis that $r_{XY} = 0$. We should bear in mind that the sampling distribution of r is approximately normal when sample size is large $(n \geq 30)$, and distributed t when the sample size is small $(n<30)$.

The t-test statistic for the significance of the correlation coefficient r is defined as:

$$t = r\sqrt{\frac{n-2}{1-r^2}} \tilde{}\, t_{1-\frac{\alpha}{2},\, n-2} \tag{11.18}$$

However, when n is sufficiently large $(n \geq 30)$, we use the standardized score for r using the Fisher z-transformation (z') to test for the significance of the correlation coefficient
r, *a method for transforming the Pearson sample to normal distribution.*

$$z' = 0.5(\ln(1+r) - \ln(1-r)) \tilde{}\, z_{\left(1-\frac{\alpha}{2}\right)} \tag{11.19}$$

The result in (11.19) is compared with the appropriate normal table.

Example 11.5:
Using the compensation data in Example 11.3,
(i)   Calculate the Pearson's correlation coefficient between X_2 and Y.
(ii) Compute the coefficient of determination in (i).
(iii) Test for the significant of the Pearson's correlation coefficient between X_2 and Y.

Solution
From the data in Example 11.3, we obtained the following values:

$$\sum X_2^2 = 921,\ \sum X_1^2 = 387,\ \sum X_1 Y = 759.61,\ \sum X_2 Y = 1165.55,\ \sum X_1 X_2 = 499,$$

$$n = 35,\ \sum X_1 = 107,\ \sum X_2 = 167,\ \sum Y = 242.76,,\ \sum Y^2 = 1690.23$$

(i) The formula for the correlation coefficient is given by:

$$r_{X_2Y} = \frac{n\sum_{i=1}^{n}X_{2i}Y_i - \left(\sum_{i=1}^{n}Y_i\right)\left(\sum_{i=1}^{n}X_{2i}\right)}{\sqrt{\left[n\sum_{i=1}^{n}X_{2i}^2 - \left(\sum_{i=1}^{n}X_{2i}\right)^2\right]\left[n\sum_{i=1}^{n}Y^2 - \left(\sum_{i=1}^{n}Y_i\right)^2\right]}}$$

Substituting for the values in the formula, we have:

$$r_{X_2Y} = \frac{35(1165.55) - (242.76)(167)}{\sqrt{\left[35(921) - (167)^2\right]\left[35\,(1690.23) - (242.76)^2\right]}} = 0.26$$

(ii) Coefficient of determination $r^2 = 0.07$
(iii) *Hypothesis*: H_0: $r_{X_2Y} = 0$ vs. H_1: $r_{X_2Y} \neq 0$

$$t = r\sqrt{\frac{n-2}{1-r^2}} = 0.26\sqrt{\frac{33}{0.93}} = 1.55$$

Critical value: $t_{0.975,\ 33} = 2.042$
Decision rule: Reject H_0 if $|t| \geq t_{0.975,\ 33}$—since 1.55<2.042, then we do not reject null hypothesis
Conclusion: The correlation coefficient between X_2 and Y are not significantly different from zero.

The R script for testing correlation coefficient for a small sample
In the code below, we will use R to obtain the correlation coefficient and confidence interval for the correlation. We compute the correlation between years of experience (*empl*) and compensation on level of education (*comp*). Our t-statistic is 1.53 with corresponding p-value of 0.14. This indicates that the correlation between years of experience (*empl*) and compensation on level of education (*comp*) is not statistically significant. In addition, at 95% CI, the true correlation between the two variables should lie between -0.08 and 0.54.

```
# Pearson's product-moment correlation using Example 11.5 compensation data
cor.test(mydata$empl, mydata$comp, method='pearson')

    Pearson's product-moment correlation

data: mydata$empl and mydata$comp
t = 1.5264, df = 33, p-value = 0.1364
alternative hypothesis: true correlation is not equal to 0
95 percent confidence interval:
 -0.08359426 0.54353702
sample estimates:
    cor
0.2568063
```

Example 11.6:
Use the following data to test for the significance of the correlation coefficient.
   $n = 50, r = 0.75$ and $\alpha = 1\%$

Solution
Hypothesis: $H_0: \rho = 0$ versus $H_1: \rho \neq 0$
Test statistic:

$$z' = 0.5(\ln(1+r) - \ln(1-r))$$

Substitute for the value of r to get z':

$$z' = 0.5*(\ln(1+0.75) - \ln(1-0.75)) = 0.973$$

Critical value: $z_{(1-\frac{\alpha}{2})} = z_{0.995} = 2.57$ (from the normal table)
Decision rule: Reject H_0 if $|z'| \geq z_{0.975}$—since 0.973<2.57, then we do not reject the null hypothesis
Conclusion: The correlation coefficient is not significantly different from zero.

The R script for testing correlation coefficient for large sample
This is a simplified version on how to compute the Fisher z-transformation (z') for the test of significance of the correlation coefficient r.
   The Fisher z-transformation (z') for r:

```
r<-0.75
z.prime<-0.5*(log(1+0.75)-log(1-0.75))
z.prime
[1] 0.9729551
```

Compute the critical value at 0.01 significance level:

```
alpha = 0.0
z.prime = qnorm(1-alpha/2)
z.prime
[1] 2.575829
```

This result is similar to the result we got above.

Exercises for Chapter 11

1. (a) Explain the concept of regression analysis.
    (b) With the aid of an example, differentiate between simple linear regression and multiple linear regression.
    (c) The table below shows data on the gross domestic product growth rate and government investment in a country between 1981 and 2016. Regress GDP growth rate on the total investment (% GDP) and interpret your result.

Year	GDP Growth rate	Total Investment (% GDP)	Year	GDP Growth rate	Total Investment (% GDP)
1981	−13.13	35.22	1999	0.47	6.99
1982	−1.05	31.95	2000	5.32	7.02
1983	−5.05	23.01	2001	4.41	7.58
1984	−2.02	14.22	2002	3.78	7.01
1985	8.32	11.97	2003	10.35	9.90
1986	−8.75	15.15	2004	33.74	7.39
1987	−10.75	13.61	2005	3.44	5.46
1988	7.54	11.87	2006	8.21	8.27
1989	6.47	11.74	2007	6.83	9.25
1990	12.77	14.25	2008	6.27	8.32
1991	−0.62	13.73	2009	6.93	12.09
1992	0.43	12.75	2010	7.84	16.56
1993	2.09	13.55	2011	4.89	15.53
1994	0.91	11.17	2012	4.28	14.16
1995	−0.31	7.07	2013	5.39	14.17
1996	4.99	7.29	2014	6.31	15.08
1997	2.80	8.36	2015	2.65	14.83
1998	2.72	8.60	2016	-1.62	12.60

2. (a) List the types of regression analysis.
   (b) What are the assumptions of simple linear regression?
   (c) The table below shows the log of floor space area (measured in sq. feet), the log of the number of bedrooms, and the log of house price (measured in $'million).
   (i) Perform a regression analysis to show the relationship between the three (3) variables. Use the log of house price as the dependent variable (Y).
   (ii) Interpret your result.
   (iii) Comment on the output.

log of floor space area (sq. feet)	8.51	8.77	8.55	8.63	7.49	8.61	8.55	8.44	8.42	8.34
log of number of bedrooms	1.61	1.39	1.39	0.69	1.10	1.61	1.39	1.39	1.39	1.61
log of house price ($'m)	0.34	0.22	0.30	0.18	0.20	0.26	0.35	0.27	0.14	0.41
log of floor space area (sq. feet)	8.39	8.52	8.69	8.60	8.40	8.39	8.25	8.71	8.55	8.55
log of number of bedrooms	1.10	1.61	1.10	0.69	1.61	1.10	1.10	1.61	1.39	1.39
log of house price ($'m)	0.17	0.24	0.22	0.22	0.35	0.19	0.10	0.48	0.43	0.35
log of floor space area (sq. feet)	8.48	8.38	8.76	8.59	8.37	8.47	8.63	8.56	8.61	8.22
log of number of bedrooms	1.61	1.10	1.10	0.69	1.10	1.10	1.10	1.10	1.61	1.39
log of house price ($'m)	0.36	0.23	0.48	0.25	0.18	0.23	0.29	0.20	0.36	0.34

3. (a) Explain the correlation coefficient.
   (b) What are the assumptions of a correlation test?
   (c) Explain the coefficient of determination.
   (d) The summary statistics of two variables are given below:

$$\sum XY = 216.24, \ \sum X = 21.20, \ \sum Y = 204.21, \ \sum X^2 = 22.90, \ \sum Y^2 = 2187.59, \ n = 20$$

   (i) Calculate the Pearson's correlation coefficient (r).
   (ii) Test for the significance of r.

4. (a) Show that the least square estimates for the simple linear regression are given as:

$$\beta_0 = \bar{Y} - \beta_1 \bar{X} \text{ and}$$

$$\beta_1 = \frac{Cov(X, Y)}{Var(X)} 3$$

   (b) Using a diagram, explain different types of correlation.

5. The strength (MPa) of steel and the diameter (mm) are shown in the table below:

Diameter (mm)	75	80	85	90	95	100	105	110	115	120
Strengtd (MPa)	250	320	350	400	435	480	500	520	550	570

   (a) Draw a scatter diagram to illustrate the information.
   (b) Calculate the least square line of regression of strength (MPa).
   (c) Interpret the model.
   (d) Estimate the strength (MPa) of the steel when the diameter (mm) is 150.

6. The following table shows the prices ($) of breads in a supermarket and the quantities sold in a week.

Price (X)	0.3	0.60	1.40	-2.10	-2.80	4.20	5.0	5.60
Quantity Sold (Y)	55	49	40	37	35	27	15	8

(a) Plot the data on a scatter diagram.
(b) Draw the regression line on your scatter diagram.
(c) From the regression equation obtained in (b), estimate the quantity to be sold if the price ($) of the bread is $6.50.
(d) Calculate Pearson's correlation coefficient between the price of bread and quantity sold.

7. The weekly number of sales in two complementary commodities (DVD players and DVD disks) are given in a table below:

Week	1	2	3	4	5	6	7	8
DVD player	20	23	18	25	15	17	16	20
DVD disks	38	39	25	28	23	26	25	30

(a) Calculate Pearson's correlation coefficient between the number of sales DVD players and DVD disks.
(b) Give the interpretation of your result in (a).
(c) Test the hypothesis that r = 0.

12 Poisson Distribution

The *Poisson distribution* was developed by the French mathematician Simeon Denis Poisson in 1837. The Poisson distribution is a discrete probability distribution. It is used to approximate the count of events that occur randomly and independently. The Poisson distribution may calculate the number of instances that an event should occur in a certain amount of time, distance, area, or volume. For instance, a random variable could be used to estimate the amount of radioactive decay in a given period of time for a certain amount of a radioactive material. If you know the rate of decay for a particular amount of the material, you can use the Poisson distribution as a good estimate of the amount of decay. We shall elaborate further on Poisson statistical properties, the derivation of the mean and the variance of a Poisson distribution, and application of the Poisson distribution. We'll demonstrate its practicality with worked examples.

12.1 Poisson Distribution and Its Properties

A discrete random variable (X) is said to be a Poisson distribution function if it has a probability distribution function of the form:

$$f(x) = \frac{\lambda^x e^{-\lambda}}{x!}, x = 0, 1, 2, \ldots \infty \tag{12.1}$$

Where λ is the shape parameter that indicates the average occurrence per interval and e is the Euler's constant, approximately 2.7183.

Figure 12.1 shows that the Poisson data skews right and the skewness is less pronounced as the lambda, λ (which is the mean of the distribution) increases. Poisson is closer to symmetric as the mean (λ) of the distribution increases.

This distribution is most applicable when we have a rare event and the number of trials is large. It implies that the rate of occurrence is small, near to zero in a large number of trials. Some examples of a Poisson distribution are the occurrence of a fatal accident within a city, the number of automobiles arriving at a traffic light in a specified period of time, the number of typographical errors in a textbook, among others.

Poisson takes values from zero to infinity, and the expected number of occurrences remains unchanged throughout an experiment. For a Poisson random variable, these two conditions should be met: (1) the number of successes

https://doi.org/10.1515/9781547401475-012

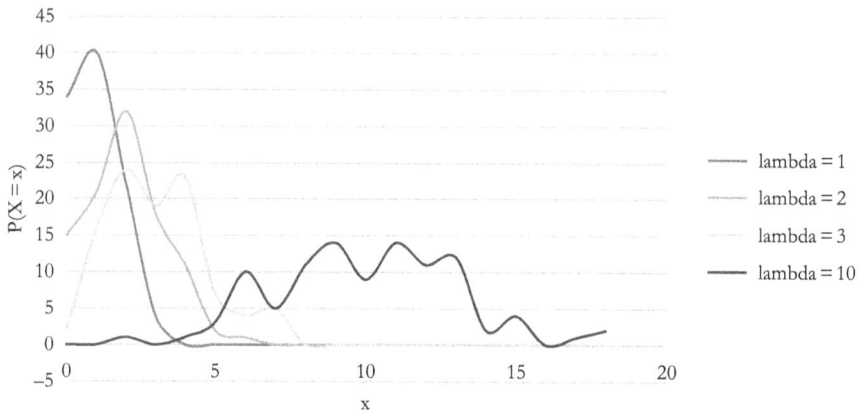

Figure 12.1: Poisson distributions at various lambdas.

in two disjointed time intervals are independent; (2) the probability of success in a short time interval is proportional to the entire length of the time interval. Alternatively, a Poisson random variable can also be applied to disjoint regions of space. Figure 12.2 shows the table of Poisson probabilities for a given value of λ and the number of successes recorded. The values in this table are computed using the formula in (12.1). This Poisson table can be read directly without computational effort. For example, suppose that a car dealer sells two exotic cars per week on average and we want to calculate the probability that in a given week, he will sell 1 exotic car. We can check the table of Poisson probabilities where mean ($\lambda = 2$) and the number of successes ($x = 1$). From the table of Poisson probabilities, the required probability is 0.2707. We get the same result by substituting for $\lambda = 2$ and $x = 1$ in the equation (12.1). Using the formula, we have:

$$f(1) = \frac{2^1 e^{-2}}{1!} = 0.270671$$

12.2 Mean and Variance of a Poisson Distribution

In this subsection, we are deriving the mean and variance of a Poisson distribution mathematically. You will see that the mean and variance of a Poisson distribution are equal.

Table of Poisson
Probabilities
For a given value of λ, entry
indicates the probability of a
specified value of X

X	0.1	0.2	0.3	0.4	0.5	0.6	0.7	0.8	0.9	1.0
0	0.9048	0.8187	0.7408	0.6703	0.6065	0.5488	0.4966	0.4493	0.4066	0.3679
1	0.0905	0.1637	0.2222	0.2681	0.3033	0.3293	0.3476	0.3595	0.3659	0.3679
2	0.0045	0.0164	0.0333	0.0536	0.0758	0.0988	0.1217	0.1438	0.1647	0.1839
3	0.0002	0.0011	0.0033	0.0072	0.0126	0.0198	0.0284	0.0383	0.0494	0.0613
4	0.0000	0.0001	0.0003	0.0007	0.0016	0.0030	0.0050	0.0077	0.0111	0.0153
5	0.0000	0.0000	0.0000	0.0001	0.0002	0.0004	0.0007	0.0012	0.0020	0.0031
6	0.0000	0.0000	0.0000	0.0000	0.0000	0.0000	0.0001	0.0002	0.0003	0.0005
7	0.0000	0.0000	0.0000	0.0000	0.0000	0.0000	0.0000	0.0000	0.0000	0.0001

X	1.1	1.2	1.3	1.4	1.5	1.6	1.7	1.8	1.9	2.0
0	0.3329	0.3012	0.2725	0.2466	0.2231	0.2019	0.1827	0.1653	0.1496	0.1353
1	0.3662	0.3614	0.3543	0.3452	0.3347	0.3230	0.3106	0.2975	0.2842	0.2707
2	0.2014	0.2169	0.2303	0.2417	0.2510	0.2584	0.2640	0.2678	0.2700	0.2707
3	0.0738	0.0867	0.0998	0.1128	0.1255	0.1378	0.1496	0.1607	0.1710	0.1804
4	0.0203	0.0260	0.0324	0.0395	0.0471	0.0551	0.0636	0.0723	0.0812	0.0902
5	0.0045	0.0062	0.0084	0.0111	0.0141	0.0176	0.0216	0.0260	0.0309	0.0361
6	0.0008	0.0012	0.0018	0.0026	0.0035	0.0047	0.0061	0.0078	0.0098	0.0120
7	0.0001	0.0002	0.0003	0.0005	0.0008	0.0011	0.0015	0.0020	0.0027	0.0034
8	0.0000	0.0000	0.0001	0.0001	0.0001	0.0002	0.0003	0.0005	0.0006	0.0009
9	0.0000	0.0000	0.0000	0.0000	0.0000	0.0000	0.0001	0.0001	0.0001	0.0002

X	2.1	2.2	2.3	2.4	2.5	2.6	2.7	2.8	2.9	3.0
0	0.1225	0.1108	0.1003	0.0907	0.0821	0.0743	0.0672	0.0608	0.0550	0.0498
1	0.2572	0.2438	0.2306	0.2177	0.2052	0.1931	0.1815	0.1703	0.1596	0.1494
2	0.2700	0.2681	0.2652	0.2613	0.2565	0.2510	0.2450	0.2384	0.2314	0.2240
3	0.1890	0.1966	0.2033	0.2090	0.2138	0.2176	0.2205	0.2225	0.2237	0.2240
4	0.0992	0.1082	0.1169	0.1254	0.1336	0.1414	0.1488	0.1557	0.1622	0.1680
5	0.0417	0.0476	0.0538	0.0602	0.0668	0.0735	0.0804	0.0872	0.0940	0.1008
6	0.0146	0.0174	0.0206	0.0241	0.0278	0.0319	0.0362	0.0407	0.0455	0.0504

Figure 12.2: Table of Poisson probabilities.

12.2.1 Mean

The following is our standard form of the mean:

$$E(x) = \sum xf(x)$$

Substituting the Poisson function we get:

$$E(x) = \sum_{x=0}^{\infty} x \frac{\lambda^x e^{-\lambda}}{x!}$$

or

$$E(x) = \sum_{x=0}^{\infty} x \frac{\lambda^{x-1} e^{-\lambda} \lambda}{x(x-1)!}$$

and

$$E(x) = \lambda \sum_{x=0}^{\infty} \frac{\lambda^{x-1} e^{-\lambda}}{(x-1)!}$$

since

$$\sum_{x=0}^{\infty} \frac{\lambda^{x-1} e^{-\lambda}}{(x-1)!} = 1$$

Therefore,

$$E(x) = \lambda \qquad (12.2)$$

So the mean of the Poisson distribution is lambda.

12.2.2 Variance

Using our standard formula for variance we have:

$$variance = E(x^2) - (E(x))^2$$

We use the following clever fact to simplify the formula for variance of the Poisson distribution in the following few steps:
From

$$x^2 = x(x-1) + x$$

Then

$$E(x^2) = E(x(x-1)) + E(x)$$

$$E(x(x-1)) = \sum_{x=0}^{\infty} x(x-1) f(x)$$

$$E(x(x-1)) = \sum_{x=0}^{\infty} x(x-1) \frac{\lambda^x e^{-\lambda}}{x!}$$

$$E(x(x-1)) = \sum_{x=0}^{\infty} x(x-1) \frac{\lambda^{x-2} e^{-\lambda} \lambda^2}{x(x-1)(x-2)!}$$

$$E(x(x-1)) = \lambda^2 \sum_{x=0}^{\infty} \frac{\lambda^{x-2} e^{-\lambda}}{(x-2)!}$$

Since

$$\sum_{x=0}^{\infty} \frac{\lambda^{x-2} e^{-\lambda}}{(x-2)!} = 1$$

Then

$$E(x(x-1)) = \lambda^2$$

And therefore,

$$var(x) = E(x(x-1)) + E(x) - (E(x))^2$$

And since E(x) = λ from our derivation of the mean above,

$$var(x) = \lambda^2 + \lambda - \lambda^2$$

Thus,

$$var(x) = \lambda \qquad (12.3)$$

The variance of the Poisson distribution is also lambda. This shows that the mean and variance of a Poisson distribution are the same.

Example 12.1:
A paper mill produces writing pads and the probability of a writing pad being defective is 0.02. If a sample of 800 writing pads is selected, what is the probability that: (i) none are defective, (ii) one is defective, (iii) two are defective, and (iv) three or more are defective.

Solution
In this example, the probability of a writing pad being defective (p) = 0.02 and the sample selected $n = 800$.
    Calculate the lambda (mean):

$$\lambda = np = 800*0.02 = 16$$

(i)   The probability that none of the selected samples are defective:

$$f(0) = \frac{16^0 e^{-16}}{0!} = e^{-16}$$

(ii)   The probability that one is defective:

$$f(1) = \frac{16^1 e^{-16}}{1!} = 16e^{-16}$$

(iii) The probability that two are defective:

$$f(2) = \frac{16^2 e^{-16}}{2!} = 128e^{-16}$$

(iv)   The probability that three or more are defective:

$$f(x \geq 3) = 1 - (f(0) - f(1) - f(2)) = 1 - e^{-16} - 16e^{-16} - 128e^{-16}$$

$$f(x \geq 3) = 1 - f(x < 3) = 1 - e^{-16} - 16e^{-16} - 128e^{-16} = 0.999$$

R code solving Example 12.1

The following are solutions to Example 12.1.

To calculate the probability of a Poisson distribution, the general format is:

```
ppois(q, lambda, lower.tail = TRUE, log.p = FALSE)

where q is the vector of quantiles
      lambda is the vector of positive means
      lower.tail is logical; if TRUE(default), probabilities are P(X ≤ x), other-
wise P(X > x)
      log.p is logical; if TRUE, probabilities P are given as log(p)
```

```
(i) The probability that none of the selected sample are defective:
n<-800
p<-0.02
lambda <-n*p
ppois(0, lambda=16, lower = TRUE)
[1] 1.125352e-07
```

This result is the same e $^{-16}$
 gotten in Example 12.1.
```
(ii) The probability that one is defective:
exact.1<-ppois(1, lambda=16)- ppois(0, lambda=16)
[1] 1.800563e-06
```

The evaluation of the quantity $16e^{-16}$
 gives 1.800563e-06.

```
(iii) The probability that two are defective:
exact.2<-ppois(2, lambda=16) - ppois(1, lambda=16)
[1] 1.44045e-05
```

The answer 1.44045e-05 is the same as $128e^{-16}$.

```
(iv) The probability that three or more are defective:
ppois(3, lambda=16, lower = FALSE)  # upper tail
[1] 0.9999069
```

The upper tail is used here because we are considering the right side of the value 3 or the value of 3 and above. Thus, we are able to use R code to solve the problem in Example 12.1.

Example 12.2:

The number of customers arriving at an eatery per minute has a Poisson with mean of 3. Assume that the number of arrivals in two different minutes are independent. Find the probability that:

(i)   no customers come in during a period of a minute.

(ii) 1 customer comes in during a period of a minute.

(iii) at least 2 customers will arrive in a given two minute period.

Solution

The mean of the Poisson distribution is 3:

(i) $\lambda = 3$

Substitute for $\lambda = 3$ and $x = 0$.

The probability that no customers come in during a period of a minute:

$$f(0) = \frac{3^0 e^{-3}}{0!} = 0.0498$$

(ii) The probability that 1 customer comes in during a period of a minute:

$$f(1) = \frac{3^1 e^{-3}}{1!} = 0.1494$$

(iii) The probability that at least 2 customers will arrive in a given two minute period:

$$f(x \geq 2) = 1 - f(0) - f(1)$$

$$f(x \geq 2) = 1 - 0.0498 - 0.1494$$

$$f(x \geq 2) = 0.8009$$

R code to calculate the probability of the Poisson distribution in Example 12.2

```
lambda =3
p0<-ppois(0, lambda=3, lower = TRUE)
[1] 0.04978707
```

The probability that no customers come in during a period of a minute is 0.0498.

```
exact.p1<- ppois(1, lambda) - ppois(0, lambda)
[1] 0.1493612
```

Despite omission of the option *"lower"* in the statement above, R assumes *lower.tail = TRUE* by default.

```
p.greater2<- 1- p0 - exact.p1
[1] 0.8008517
```

The probability that at least 2 customers will arrive in a given two minute period is 0.8009. We got similar results with Example 12.2.

Example 12.3:
An operations manager has the option of using one of two machines (A or B) to produce a particular product. The energy output of each machine is represented well by a Poisson distribution with machine A having a mean of 8.25 and machine B having a mean of 7.50. Assuming that the energy input remains the same, the efficiency of machine A is $f(x) = 2x^2 - 8x + 6$ and the efficiency of machine B is $f(y) = y^2 + 2y + 1$. Which of the machines has the better expected efficiency?

Solution
Machine A's expected efficiency is:

$$E(x) = E(2x^2 - 8x + 6)$$

$$E(x) = 2E(x^2) - 8E(x) + E(6)$$

From the general equation:

$$Var(x) = E(x^2) - (E(x))^2$$

This implies that,

$$E(x^2) = Var(x) + (E(x))^2$$

Substitute for $E(x^2)$ to get:

$$E(x) = 2\left[Var(x) + (E(x))^2\right] - 8E(x) + 6$$

Also, substitute for the mean and variance of the distribution to get:

$$E(x) = 2\left[8.25 + 8.25^2\right] - (8 \times 8.25) + 6 = 92.63$$

Machine B's expected efficiency is:

$$E(y) = E(y^2 + 2y + 1)$$

$$E(y) = E(y^2) + 2E(y) + E(1)$$

$$E(y) = \left[Var(y) + E(y)^2\right] + 2E(y) + 1$$

Since

$$E(y^2) = V(y) + E(y)^2$$

Therefore,

$$E(y) = [7.50 + 7.50^2] + (2 \times 7.50) + 1 = 79.75$$

So Machine A is more efficient than Machine B.

R code for solving Example 12.3

In a Poisson distribution, the mean and variance are equal, therefore the mean of machine A equals the variance of machine A = 8.25.

```
# compute efficiency for machine A
mean.x<-8.25
var.x<-8.25
```

Referring to this, $E(x) = 2\left[Var(x) + (E(x))^2\right] - 8E(x) + 6$, we got efficiency for the machine A as *eff.x*.

```
eff.x<-2*(var.x+mean.x**2)-8*mean.x+6
[1] 92.625
```

Also, the mean of machine B equals the variance of machine B = 7.50, and then computing the efficiency for machine B:

```
# compute efficiency for machine B
mean.y<-7.50   # expected efficiency
var.y<-7.50    # variance efficiency
eff.y<- var.y+mean.y**2+2*mean.y+1
[1] 79.75
```

The efficiency of machine A is 92.63 and machine B is 79.75, so machine A is more efficient.

12.3 Application of the Poisson Distribution

The Poisson distribution has applications in various aspects of life. In health care, the Poisson distribution is helpful in birth defect detection and genetic mutations. It also helps to model the occurrence of rare diseases, such as leukemia. In the area of transport, Poisson is good at accounting and predicting the occurrence of highway accidents, and monitoring the number of automobiles arriving at a traffic light within a period of time. Poisson can also be applied in the telecommunications and communications sectors, by determining the number of calls received by a customer care center in a particular minute and the number of network failures per day. In the area of production, it can be used to capture the failure of a machine in a particular month and the number of defective products in a production line.

12.4 Poisson to Approximate the Binomial

The Poisson distribution is a special case of the binomial distribution and Poisson is useful in handling rare events. In the event that the binomial random variable has an extremely large number of trials (n) and the probability of success (p) is very small, then the Poisson distribution provides a good approximation of the binomial distribution when $n \geq 100$, and $np \leq 10$.

Proof
The Poisson (λ) is an approximation to the Binomial (n, p) for a large n, small p, and $\lambda = np$.
Then

$$p = \frac{\lambda}{n}$$

Substitute for p into the binomial distribution and then take the limit as n tends to infinity:

$$\lim_{n \to \infty} P(X = k) = \lim_{n \to \infty} \frac{n!}{(n-k)!k!} \left(\frac{\lambda}{n}\right)^k \left(1 - \frac{\lambda}{n}\right)^{n-k} \tag{12.4}$$

Take out the constant terms in (12.4):

$$\lim_{n \to \infty} P(X = k) = \frac{\lambda^k}{k!} \lim_{n \to \infty} \frac{n!}{(n-k)!} \left(\frac{1}{n}\right)^k \left(1 - \frac{\lambda}{n}\right)^n \left(1 - \frac{\lambda}{n}\right)^{-k} \tag{12.5}$$

We can take the limit of the RHS one term after the other in (12.5):

$$\lim_{n \to \infty} \frac{n!}{(n-k)!} \left(\frac{1}{n}\right)^k$$

$$\lim_{n \to \infty} \frac{n(n-1)(n-2)\ldots(n-k)(n-k-1)}{(n-k)(n-k-1)\ldots(1)} \left(\frac{1}{n}\right)^k$$

$$\lim_{n\to\infty} \frac{n(n-1)(n-2)\ldots(n-k+1)}{n^k} \tag{12.6}$$

As $n \to \infty$, then k terms tend to 1.

The equation (12.6) can be written as:

$$\lim_{n\to\infty} \frac{n(n-1)(n-2)\ldots(n-k+1)}{n^k} \left(\frac{n}{n}\right)\left(\frac{n-1}{n}\right)\left(\frac{n-2}{n}\right)\ldots\left(\frac{n-k+1}{n}\right) \tag{12.7}$$

The second step is to take the limit of the middle term in (12.5):

$$\lim_{n\to\infty} \left(1-\frac{\lambda}{n}\right)^n \tag{12.8}$$

Since e = 2.718, then

$$e = \lim_{x\to\infty} \left(1-\frac{1}{x}\right)^x \tag{12.9}$$

Let $x = -\frac{n}{\lambda}$

Substitute for x into (12.8), then we have:

$$\lim_{n\to\infty} \left(1-\frac{\lambda}{n}\right)^n = \lim_{x\to\infty} \left(1-\frac{1}{x}\right)^{x(-\lambda)} = e^{-\lambda} \tag{12.10}$$

Take the limit of the last term of the RHS of (12.5)

$$\lim_{n\to\infty} \left(1-\frac{\lambda}{n}\right)^{-k} \tag{12.11}$$

As $n \to \infty$, then $(1)^{-k}$ tends to 1.

Putting all the results (12.7), (12.10), and (12.11) together into (12.5), we have:

$$\lim_{n\to\infty} P(X=k) = \frac{\lambda^k}{k!} \lim_{n\to\infty} \frac{n!}{(n-k)!} \left(\frac{1}{n}\right)^k \left(1-\frac{\lambda}{n}\right)^n \left(1-\frac{\lambda}{n}\right)^{-k} = \left(\frac{\lambda^k}{k!}\right)(1)\left(e^{-\lambda}\right)(1) \tag{12.12}$$

Therefore,

$$P(\lambda, k) = \left(\frac{\lambda^k e^{-\lambda}}{k!}\right)$$

This gives the Poisson distribution with k successes per period given the parameter λ.

Example 12.4:

Suppose a random variable X has a binomial distribution with $n = 120$ and $p = 0.01$, use the Poisson distribution to calculate the following: (i) P (X = 0), (ii) P (X = 1), (iii) P (X = 2), and (iv) P (X > 2).

Solution

(i) $\lambda = np = 1.2$

$$P(X=0) = \frac{1.2^0 e^{-1.2}}{0!} = 0.3012$$

(ii) $P(X=1) = \frac{1.2^1 e^{-1.2}}{1!} = 0.3614$

(iii) $P(X=2) = \frac{1.2^2 e^{-1.2}}{2!} = 0.2169$

(iv) $P(X>2) = 1 - P(X \le 2) = 1 - P(X=0) - P(X=1) - P(X=2) = 0.1205$

R code for solutions to Example 12.4

```
# Question 12.4i
n<-120
p<-0.01
lambda = n*p
p0<-ppois(0, lambda, lower = TRUE)
[1] 0.3011942
```

The probability that $P(X=0)$ is 0.3012.

```
# Question 12.4ii
p1<-ppois(1, lambda, lower = TRUE) - p0
[1] 0.3614331
```

The probability that $P(X=1)$ is 0.3614.

```
# Question 12.4iii
p2<-ppois(2, lambda)-ppois(1, lambda)
p2
[1] 0.2168598
```

The probability that $P(X=2)$ is 0.2169.

```
# Question 12.4iv
p.greater2<-1-ppois(2, lambda, lower=TRUE)
p.greater2
[1] 0.1205129
```

The probability that $P(X > 2)$ is 0.1205.

Exercises for Chapter 12

1. (a) Define the Poisson distribution and its properties.
   (b) Show that the mean and the variance of a Poisson distribution are equal.
   (c) A production manager takes a sample of twenty-five (25) textbooks and examines them for the number of defectives pages. The outcome of his findings is summarized in the table below.

Number of defects	0	1	2	3	4	5
Frequency	10	4	3	2	3	3

   Find the probability of finding a textbook chosen at random that contains 2 or more defective pages.

2. (a) Show that the Poisson (θ) is an approximation to the binomial (n, p) as $n \to \infty$, and $p \to 0$.
   (b) Consider a random variable X that has a binomial distribution with $n = 100$ and $p = 0.005$, use the Poisson distribution to calculate the following: (i) P (X = 0), (ii) P (X = 1), and (iii) P (X > 1).

3. (a) Explain the areas of interest where the Poisson process can be applied.
   (b) The ABC company supplies a supermarket with groceries weekly. Experience has shown that 3% of the total supply of the groceries in a given week are defective. A manager of the supermarket decided to check the number of defective items for a particular grocery and took a sample of 105 items. What is the probability that: (i) none of the products is defective? (ii) one of the products is defective? (iii) 2 or more of the products are defective?

4. A courier company delivers 200 parcels every 30 days and the management of the company observed that most of the deliveries take place between 11:00 a.m. and 1:00 p.m because of less traffic on the road. The management decided to make as much staff available during this period of the day. Use the Poisson distribution to find the probability of delivering: (i) no

parcel, (ii) one parcel, (iii) two peacels, and (iv) three percels in that period of the day. (v) How many days within a 30-day period would a delivery of four or more parcels be expected?

5. An insurance company sells a life insurance policy to people that have attained age 50 and above. The actuarial probability that somebody at age 50 or above will die within one year of the policy is 0.0008. If the life insurance policy is sold to 7,500 people of the same age group, there is a possibility that 6 people at 50 years and above will die in the within the year. What is the probability that the insurance company will pay exactly 6 claims on the 7,500 policies sold in the next year?

6. A car dealer claimed that he sold an average of 3 new cars per week. Assuming the sales follow a Poisson distribution, what is the probability that: (i) he will sell exactly 3 new cars, (ii) fewer than 3 new cars, and (iii) more than 3 new cars in a given week?

13 Uniform Distributions

The uniform distribution is used when each outcome has the same probability of occurring. A typical example of a uniform distribution is rolling a fair die where each of the six numbers has equal probability of occurring. The uniform distribution has constant probability across all values and the uniform distribution can be either a discrete or continuous distribution. In this section, we are considering a continous uniform distribution. We are deriving the mean and variance of a continuous uniform distribution, and we shall illustrate these with examples.

13.1 Uniform Distribution and Its Properties

If X is a continuous random variable, then X is said to have a uniform distribution over the interval [a, b], if its probability density function is defined as:

$$f(x) = \frac{1}{b-a}, a \leq x \leq b$$

The uniform distribution is denoted as $X \sim U(a, b)$. The uniform distribution is also known as a *rectangular distribution*.

The rectangle in Figure 13.1 shows that the length of the base is $(b\text{-}a)$ and the height of the rectangle is $\frac{1}{b-a}$. The total area under the curve of this probability distribution function is the product of the height of the rectangle and the length of the base; thus, the total area is 1. The area under $f(x)$ and between the points a and b is expected to be 1 and $f(x) > 0$, therefore, $f(x)$ is a probability density function. There are an infinite number of possible values of a and b; thus, there are an infinite number of possible uniform distributions. The commonly used continuous uniform distribution is where $a = 0$ and $b = 1$. A uniform distribution is useful when every variable has and equal or the exact same chance of happening.

13.2 Mean and Variance of a Uniform Distribution

We derive the mean and the variance of a uniform distribution in this subsection.

https://doi.org/10.1515/9781547401475-013

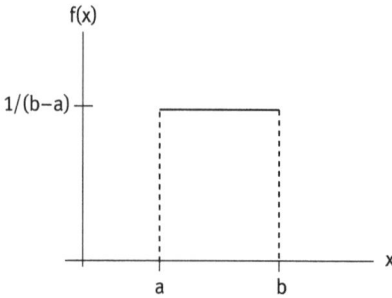

Figure 13.1: A uniform distribution.

13.2.1 Mean

The expected value of X is

$$E(x) = \int_a^b xf(x)dx$$

Substituting for $f(x)$ yields:

$$E(x) = \int_a^b \frac{x}{b-a}dx$$

Solving the integral yields:

$$E(x) = \left[\frac{x^2}{2(b-a)}\right]_a^b$$

$$E(x) = \left[\frac{b^2}{2(b-a)} - \frac{a^2}{2(b-a)}\right] = \frac{b^2-a^2}{2(b-a)}$$

$$E(x) = \frac{(b+a)(b-a)}{2(b-a)} = \frac{(b+a)}{2}$$

So the mean is $\dfrac{(b+a)}{2}$ and is the average of the two limits (a and b).

13.2.2 Variance

The variance of a uniform distribution is derived as follows:

$$Var(x) = E(x^2) - (E(x))^2$$

$$E(x^2) = \int_a^b x^2 f(x) dx$$

$$E(x^2) = \int_a^b x^2 \frac{1}{b-a} dx$$

$$E(x^2) = \frac{1}{b-a} \int_a^b x^2 dx$$

Solving the integral yields:

$$E(x^2) = \frac{1}{b-a} \left[\frac{x^3}{3} \right]_a^b$$

$$E(x^2) = \frac{1}{3(b-a)} \left[x^3 \right]_a^b$$

$$E(x^2) = \frac{1}{3(b-a)} \left[b^3 - a^3 \right]$$

Since $[b^3 - a^3] = (b^2 + ab + a^2)(b - a)$

Then $\qquad E(x^2) = \frac{1}{3(b-a)} \left[(b^2 + ab + a^2)(b - a) \right]$

$$E(x^2) = \frac{1}{3} \left[(b^2 + ab + a^2) \right] \text{ since } Var(x) = E(x^2) - (E(x))^2$$

Then $\qquad Var(x) = \frac{1}{3} \left[(b^2 + ab + a^2) \right] - \left(\frac{(b+a)}{2} \right)^2$

$$Var(x) = \frac{(b^2 + ab + a^2)}{3} - \frac{(b+a)^2}{4}$$

$$Var(x) = \frac{4(b^2 + ab + a^2) - 3(b+a)^2}{12}$$

$$Var(x) = \frac{4b^2 + 4ab + 4a^2 - 3b^2 - 6ab - 3a^2}{12}$$

$$Var(x) = \frac{b^2 - 2ab + a^2}{12}$$

$$Var(x) = \frac{(b-a)^2}{12}$$

Hence, the variance of a uniform distribution is $\frac{(b-a)^2}{12}$, the square of the difference of points a and b, divided by 12.

Example 13.1:

John arrived at his office building and wanted to take the elevator to the 10th floor where his office is located. Due to the congestion at the building, it takes between 0 and 60 seconds for the elevator to arrive at the ground floor. Assume that the arrival of the elevator to the ground floor is uniformly distributed: (i) What is the probability that the elevator takes less than 15 seconds to arrive at the ground floor? (ii) Find the mean of the arrival of the elevator at the ground floor? (iii) Find the standard deviation of the arrival of the elevator at the ground floor?

Solution

(i) Let the intervals in seconds be a = 0, b = 60, and c =15.

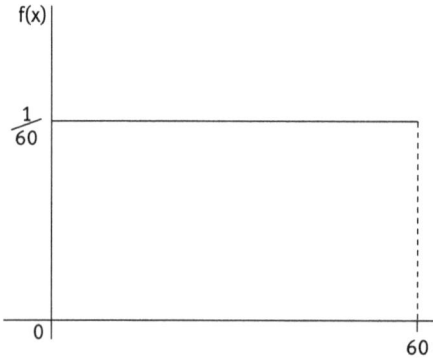

$$P(0 \leq x \leq 15) = \int_0^{15} \frac{1}{b-a} dx$$

$$P(0 \leq x \leq 15) = \int_0^{15} \frac{1}{60-0} dx$$

$$P(0 \leq x \leq 15) = \left[\frac{x}{60}\right]_0^{15}$$

$$P(0 \leq x \leq 15) = \left[\frac{15-0}{60}\right] = 0.25$$

The probability that the elevator will arrive at the ground floor in 15 seconds is one-fourth of the time.

(ii) $mean = E(x) = \dfrac{(b+a)}{2}$

$$E(x) = \frac{(60+0)}{2} = 30 seconds$$

The mean arrival time is 30 seconds.

(iii) $Var(x) = \dfrac{(b-a)^2}{12} = \dfrac{(60-0)^2}{12} = \dfrac{60^2}{12}$

$$SD(x) = \sqrt{\frac{60^2}{12}} = \frac{60}{\sqrt{12}} = 17.32 \text{ seconds}$$

The standard deviation is just over 17 seconds.

R code of a uniform distribution

The distribution function of a uniform distribution in R is generated by *punif*. The format is:

```
punif(q, min = 0, max = 1, lower.tail = TRUE, log.p = FALSE)
```

where q is the vector of quantiles

min is the lower limit of the distribution

max is the upper limit of the distribution

lower.tail is logical; if TRUE (default), probabilities are $P[X \leq x]$, otherwise, $P[X > x]$

log.p is logical; if TRUE, probabilities p are given as $log(p)$

The R code that provides the solutions to Example 13.1 is:

```
# Solution to example 13.1(i)
punif(15,min=0,max=60)
[1] 0.25
```

The probability that the elevator will arrive in 15 seconds is 0.25

```
# Solution to example 13.1(ii)
min<-0
max<-60
punif(15,min=0,max=60)
mean.x<-(max+min)/2
mean.x
[1] 30
```

The mean arrival time is 30 seconds.

```
# Solution to example 13.1(iii)
var.x<-((max-min)**2)/12
sd.x<-sqrt(var.x)
sd.x
[1] 17.32051
```

The standard deviation is about 17 seconds.

Example 13.2:

A meeting of 10 department heads to give an update of their stewardship in their respective departments and the managing director gives each department head 5 minutes to brief the management about new developments in their department.. (i) Find the probability that the department heads will finish their brief within 3 mins 30 seconds. (ii) How many of the department heads can finish their brief within 3 minutes 30 seconds?

Solution

Let's convert all time intervals to seconds:

Interval (in seconds) of probability distribution = [0, 300]

$$f(x) = \frac{1}{b-a} = \frac{1}{300}$$

$$P(0 \leq x \leq 210) = \int_0^{210} \frac{1}{300} dx$$

$$P(0 \leq x \leq 210) = \left[\frac{x}{300}\right]_0^{210}$$

$$P(0 \leq x \leq 210) = \left[\frac{210}{300}\right] = 0.7$$

Therefore, the probability that the department heads will finish their talk within 3 minutes 30 seconds is 0.7.

The number of department heads that can finish up the talk within 3 minutes 30 seconds is: $0.7 \times 10 = 7$ department heads.

R code for the solution to Example 13.2

```
# Solution to Example 13.2(i)
p.210<-punif(210,min=0,max=300)
p.210
[1] 0.7
```

The possibility that the department heads will finish their talk within 3 minutes 30 seconds is 0.7.

```
# Solution to Example 13.2(ii)
n.heads<- p.210*10
n.heads
[1] 7
```

Thus, 7 department heads can finish up their brief within 3 minutes 30 seconds.

Example 13.3:

Two managers, Peter and Paul, agree to meet for an after-work business meeting. Both of them randomly arrive between 5:45 p.m. and 6:00 p.m. What is the probability that Peter arrives at the venue at least 5 minutes before Paul?

Solution

Let x be the arrival time for Peter and y be the arrival time for Paul.

Time interval = 6: 00 p.m. – 5:45 p.m. = 15 minutes

Time interval of probability distribution = [0, 15]

$$f(x) = \frac{1}{b-a} = \frac{1}{15}$$

$$P(x \geq 5) = \int_{5}^{15} \frac{1}{15} dx$$

$$P(x \geq 5) = \left[\frac{x}{15} \right]_{5}^{15}$$

$$P(x \geq 5) = \frac{15}{15} - \frac{5}{15} = \frac{10}{15} = 0.67$$

The probability that Peter arrives at the venue at least 5 minutes before Paul is 0.67.

R code for the solution to Example 13.3

```
# Solution to Example 13.3
p.greater5<-1-punif(5,min=0,max=15)
p.greater5
[1] 0.6666667
This result (0.67) gives the probability of Peter arriving the venue at least 5
minutess before Paul.
```

Example 13.4:

A gas station manager claims that the minimum volume of gasoline sold per day is 5,000 liters and the maximum sold per day is 6,500 liters. Assume that the service at the gas station is a uniform distribution. Find the probability that the volume of the product sold per day will fall between 5,500 and 6,200 liters?

Solution

Given $a = 5,000$ and $b = 6,500$

$$f(x) = \frac{1}{b-a} = \frac{1}{1500}$$

$$P(5500 \leq x \leq 6200) = \int_{5500}^{6200} \frac{1}{1500} dx$$

$$P(5500 \leq x \leq 6500) = \left[\frac{x}{1500} \right]_{5500}^{6200}$$

$$P(5500 \leq x \leq 6500) = \frac{6200}{1500} - \frac{5500}{1500} = 0.47$$

R code for solution to Example 13.4

```
# Solution to Example 13.4
p.6200<-punif(6200,min=5000,max=6500)
[1] 0.8
```

The probability that 6,200 liters of gas are sold per day is 0.8.

```
p.5500<-punif(5500,min=5000,max=6500)
[1] 0.3333333
The probability that 5,500 liters of gas are sold per day is 0.33.

btw5500_6200<-p.6200-p.5500
btw5500_6200
[1] 0.4666667
```

The probability that the volume of gas sold per day will fall between 5,500 and 6,200 liters is 0.47; this is the difference in the probability of the sale of 6,200 liters and 5,500 liters.

Exercises for Chapter 13

1. (a) Define the continous uniform distribution.
    (b) Suppose X has a uniform distribution on the interval [10, 50]. That is, $X \sim U(10, 50)$. Find the probability that: (i) $P(X \leq 25)$, (ii) $P(20 \leq X \leq 32)$, and (iii) $P(X \geq 32)$.
2. (a) Let $X \sim U(\alpha, \beta)$ and show that the mean of a uniform distribution is $\frac{(\beta + \alpha)}{2}$ and variance $\frac{(\beta - \alpha)^2}{12}$.
    (b) If $f(x) = 3x^2$, $0 \leq x \leq 2$, find the expected value of X.
3. The waiting time for a bus to arrive at a bus stop is 10 minutes. Find the probability that a bus will come within 7 minutes of waiting at the bus stop, assume that the waiting time is uniformly distributed.

4. A recruitment agency conducted interviews for 20 job seekers and scheduled 5 minutes for each of the job seekers to express themselves on what changes and contributions they would bring to the company. Find the probability that: (a) the job seekers finish *within* 3 minutes, (b) the job seekers finish *after* 3 minutes, and (c) how many of these job seekers express themselves in more than 3 minutes.

5. Suppose $X \sim U(2,10)$.
   (a) What is the probability density function $f(x)$ of X?
   (b) Sketch the probability density function $f(x)$ of X.
   (c) What is $P(5 \leq X \leq 10)$?
   (d) What is $E(2X^2 - X)$?

14 Time Series Analysis

Time series analysis is the systematic approach explored to address the mathematical and statistical issues created by time correlations. Changes take place through time. The purchasing power of money is not constant. The value of the US dollar today is not the same as a dollar a month ago. People believe that changes in currency value are caused by inflation. However, the value of a dollar last year might be lower than the value of a dollar today due to other economic factors. For instance, if a dollar today can be invested or put in the bank or lent to someone, the investor or bank depositor or borrower has to be compensated for giving someone else his dollar for a year. Therefore, a dollar today will be equivalent to a dollar plus some amount a year from now (the earnings for investing the dollar for one year), thus making one dollar today worth more than a dollar a year ago. This accumulated value is then reduced by inflation.

One method we explore to assess the value of money across years is the the consumer price index (CPI), the most common time series data known. This index identifies a base year for which the value is considered to be 100. This chapter provides deeper insights on time series analysis with various examples to explain concepts used in the analysis of time series data.

14.1 Concept of Time Series Data

A *time series* exhibits a steady tendency of increase or decrease over time. Such a tendency is denoted as a trend. By plotting observations against time, conclusions can be made on a straight line whether the series experiences a rise or a fall over time. In this case, the least square approach would be used to estimate the parameters of a straight-line model. Time series data exhibits a random error that makes it difficult to identify the pattern in the dataset. In other words, a steady tendency does not mean that patterns will follow predictable patterns, only that overall, a trend can be observed and predicted, given the unavoidable occurrence of random errors.

14.1.1 Uses and Application of Time Series Analysis

Time series analysis can be used in scientific applications where major time series problems may occur. For instance in the field of economics, daily stock

https://doi.org/10.1515/9781547401475-014

market quotations or monthly unemployment figures are exposed to time series issues. Similarly, this occurs in the social sciences in terms of population series like birth rates or school enrollment. Epidemiologists may want to observe the number of influenza cases over time; a medical expert might be interested in studying blood pressure measurement over some period in order to evaluate the effectiveness of drugs taken in treating hypertension.

The initial phase in any time series research focuses on careful scrutiny of the observed data plotted over time. This method of analysis, as well as statistics, will be of great importance in summarizing the information in the data.

It is noteworthy that there are two separate approaches to time series analysis. These approaches are not mutually exclusive. These are generally identified as the *time domain approach* and the *frequency domain approach*.

14.1.1.1 Time Domain Approach

This approach is based on the presumption that correlation between adjacent points in time is best examined in terms of dependence of the current value on past values. In addition, it models some future value of a time series as a parametric function of the current and past values. The approach commences with linear regressions of the present value of a time series on its own past values as well as the past values of other series. The outcomes of this model allow its use as a forecasting tool. This modeling technique is particularly common in the field of economics. Examples of the time domain method are autoregressive integrated moving average (ARIMA) models. This approach entails more than one input series through multivariate ARIMA or with transfer function modeling (analysis of relationships between inputs and outputs of a series using a polynominal ratio). The transfer function model is a form of a very large class of models, which include the univariate ARIMA model as a special case.

The common attribute of these models is that they are multiplicative models, implying that the observed data are assumed to be generated from products of factors relating to differential or difference equation operators in response to a white noise input. Using the additive models on the same problems is a more recent technique, in which the observed data is assumed to proceed from sums of series, each linked to a unique time series structure; for instance, in economics, a series is assumed to be generated from the summation of trend, a seasonal effect, and error. This disaggregation is done with the use of the celebrated Kalman filters (equations providing recursive estimation of a process) and smoothers (to uncover hidden variables), built basically for estimation and control in space applications.

14.1.1.2 Frequency Domain Approach

This approach works on the assumption that the basic features of interest in time series analyses associate with periodic or systematic sinusoidal variations observed naturally in most data. These are curves describing a periodic smoothing in oscillations. Biological, physical, or environmental factors are usually responsible for these periodic fluctuations. For instance, sea surface temperature caused by El Niño oscillations may influence the amount of fish in the ocean. This approach has been extended to economics and social sciences, in which one might have interest in the annual periodicity of such series as monthly unemployment or monthly birth rates.

A spectral analysis is conducted to separately evaluate the variance associated with each periodicity of interest. This variance profile over frequency is regarded as the *power spectrum,* the distribution of power into frequency components, denoted as *Sxx(f).* There is no break between time domain and frequency domain methodologies, but different researchers advocate one or the other approach to analyzing data. Both approaches often produce the same results for long series, but the relative performance over short samples is better carried out in the time domain.

The best way to analyze many datasets is to utilize the two approaches in a complementary fashion. This book will provide how this can be done.

The next sections are aimed at providing a unified and reasonably complete exposition of statistical techniques used in time series analysis.

14.2 Univariate Time Series Model

How rapid is oil price changing? What will the effect be in the coming years? These questions can be answered by following a phenomenon (oil price) over time. This section will focus on investigating observations that have been reported at regularly spaced time intervals for a given span of time. These observations can be organized into a time series of the form $X_1, X_2, \ldots\ldots, X_t, \ldots, X_T$ where X_t is the value of X at time t and T is the total number of observations in the series.

The nottem time series encompasses the average monthly temperature at Nottingham from 1920 to 1939. There are 240 observations—one for each month over 20 years. This is a time series dataset built in R and it is illustrated in Figure 14.1.

The analysis of time-series data has two basic questions: What occurred (description), and what will occur next (forecasting). For the nottem data, the following questions might be asked:

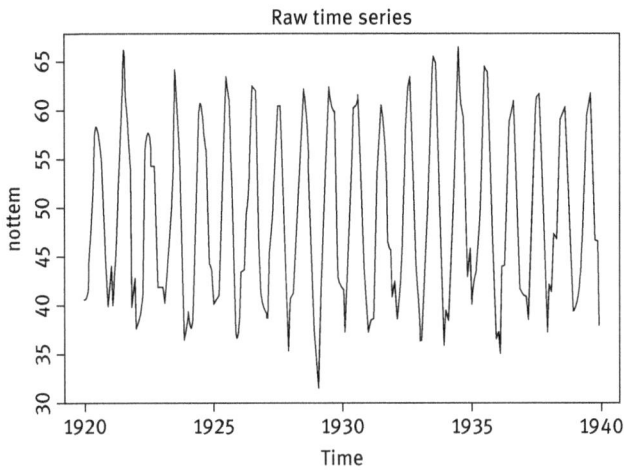

Figure 14.1: Nottem time series.

- Is the average monthly temperature changing over time?
- Are there monthly impacts, with temperature increasing and dropping in a regular pattern during these periods?
- Is it possible to forecast what the future temperature will be and, if so, to what degree of accuracy?

The ability to have an accurate prediction of average monthly temperature is critically important for the aviation industry and for human ecosystems. Forecasting the future value of a time series is a basic human activity and investigations of time series observations are applicable to real-world situations. For instance, practitioners such as economists and financial analysts use time-series data in order to understand and forecast what will occur in the financial markets. Similarly, urban planners use time-series observations to anticipate future transportation demands, and business organizations explore time-series information to forecast demand and future sales.

There are different techniques available to describe time-series data as well as forecast future values. However, R provides the most comprehensive analytic capabilities available anywhere. Therefore, this section applies the R functions to produce the most common descriptive and forecasting techniques. The functions are itemized in Table 14.1 based on their order of appearance.

The analysis of the time-series data will be explored using the command lists in Table 14.1.

Table 14.1: Commands for time series analysis.

Command	Application
ts ()	Creates a time-series object
plot ()	Graphs a time series
start()	Returns the commencement time of a time series
end ()	Returns the ending time of a time series
frequency ()	Returns the duration of a time-series
window ()	Subsets a time-series object
ma ()	Creates a simple moving-average model
stl ()	Decomposes a time series into seasonal, trend, and irregular components using LOESS
monthplot ()	Graphs the seasonal parts of a time series
Seasonplot ()	Creates a season plot
HoltWinters ()	Fits an exponential smoothing model
Forecast ()	Predicts future values of a time series
Accuracy ()	Reports fit measures for a time-series model
Ets ()	Fits an exponential smoothing model as well as the ability to automate the selection of a model
Lag ()	Creates a lagged series of a time series
Acf()	Estimates the autocorrelation function
Pacf ()	Estimates the partial autocorrelation function
Diff ()	Generates lagged and iterated differences
Ndiffs ()	Calculates the level of differencing required to remove trends in a time series
Adf.test ()	Computes an Augmented Dickey-Fuller test for a stationarity of a time series
Arima ()	Establishes autoregressive integrated moving-average models
Box.test ()	Computes a Ljung-Box test that the residuals of a time series are not dependent
Bds.test ()	Generates the BDS test that a series includes independent, identically distributed random variables
Auto.arima ()	Automates the selection of an ARIMA model

Table 14.2: Datasets (built in R) used.

Time Series	Description
Nottem	Average monthly temperature at Nottingham 1920–1939
BJ sales	Sales data without leading indicators
Austres	Quarterly time series of the number of Australian residents
BJ sales.lead	Sales data with leading indicators

The analysis will commence with techniques for creating and manipulating time series, describing and plotting them, as well as decomposing the time-series data into level, trend, seasonal, and irregular (error) components. It proceeds to the forecasting section that employs weighted averages of time-series values to anticipate future values. In this section, a set of forecasting techniques known as autoregressive integrated moving averages (ARIMA) models that utilize correlations among recent data points and among recent prediction errors to make future anticipations will be examined. In addition, the fit of models as well as the accuracy of their predictions will be assessed using the different techniques. We conclude with the provision of available resources to gain more understanding about these topics.

14.2.1 Generating a Time-Series Object in R

In order to explore R in the time series analysis, observations need to be placed into a time-series object (an R structure that requires the observations, the starting and ending time of the series, and information about its periodicity—for instance, daily, weekly, monthly, quarterly, or annual data). Having the data in a time-series object, you can utilize a variety of commands to manipulate, model, and plot the data.

With the use of *ts( )* functions, a vector of numbers or a column in a dataframe can be saved. The command for this as follows:

$$myseries <- ts\ (data,\ start =,\ end =,\ frequency =)$$

Where *myseries* refers to the time-series object, *data* represents a numeric vector entailing the observations, *start* indicates the series start time, *end* denotes the optional end time, and *frequency* is the number of observations per unit time (for example, frequency = 12 for monthly data, frequency = 4 for quarterly data, and frequency = 1 for annual data).

Example 14.1:

Consider that the data consists of monthly sale figures for two years, commencing in January 2015.

(i) Creating a time-series object

```
sales<-c(20, 25, 33, 42, 7, 34, 26, 21, 24, 13, 14, 15, 32, 18, 33, 29, 40, 54,
43, 35, 31, 27, 33, 53)
tsales<-ts(sales, start=c(2015,1), frequency=12)
tsales
        Jan Feb Mar Apr May Jun Jul Aug Sep Oct Nov Dec
2015    20 25 33 42 7 34 26 21 24 13 14 15
2016    32  18 33 29 40 54 43 35 31 27 33 53
plot(tsales)
```

To get more information about *tsales*, the following code is written:

```
start(tsales)
[1] 2015 1
end(tsales)
[1] 2016 12
frequency(tsales)
[1] 12
```

The code below allows you to create a subset with *tsales* data:

```
tsales.subset<-window(tsales, start=c(2015,3), end=c(2016,4))
tsales.subset
     Jan Feb  Mar  Apr  May  Jun Jul  Aug Sep Oct Nov Dec
2015          33   42   7    34   26  21   24  13  14  15
2016  32  18  33   29
```

14.2.2 Smoothing and Seasonal Decomposition

It is more important to describe a time series numerically and visually before taking steps to build complex models. Therefore, this section will focus on smoothing a time series in order to clarify its general trend, and decomposing the series in order to observe any seasonal influences.

14.2.2.1 Smoothing with Simple Moving Averages

The first phase in examining a time series is to graph it as indicated in Figure 14.1 (for instance, the BJ sales time series that captures the annual sales). A plot of this series is depicted in Figure 14.2. The time series exhibits an upward trend with a great deal of variation from year to year.

In order to assess any patterns in the data, you need to plot a smoothed curve that mitigates the fluctuations. The use of simple moving averages is among the most basic techniques to smooth a time series. For example, the mean of an observation and one observation before and after it, can be used instead of each data point. This method is known as a *centered moving average*. A centered moving average is expressed as:

$$S_t = \left(Y_{t-q} + \ldots + Y_t + \ldots + Y_{t+q} \right) / (2q + 1)$$

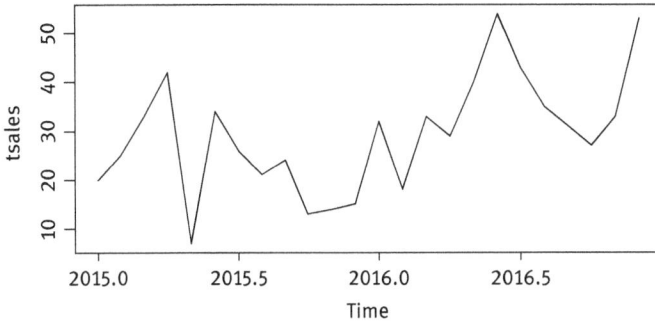

Figure 14.2: Generated sales time series.

where S_t denotes the smoothed value at time t and $k = 2q+1$ represents the number of observations that are averaged. The value of k is often selected to be an odd number. With the use of a centered moving average, the $(k-1)/2$ observations at each end of the series are lost.

There are many functions in R that can be used to generate a simple moving average. These include *SMA ( )* in the TRR package, *rollmean( )* in the zoo package, and *ma( )* in the forecast package. In order to smoothen the BJsales series, the *ma( )* function will be used since it comes with the base R installation. The following R commands are used to generate the nottem data and the 3-month, 7-month, and 11-month moving averages for nottem data are generated respectively, as shown in Figures 14.3–14.6.

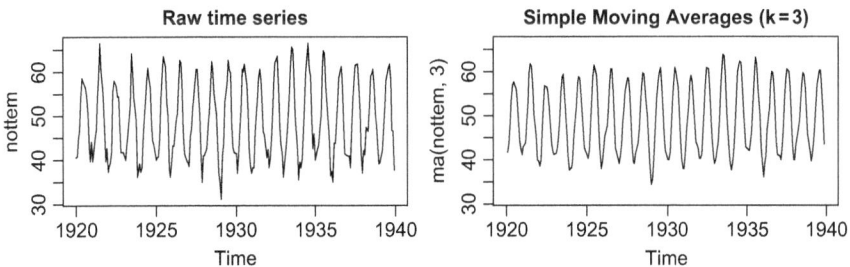

Figure 14.3: Moving average (k=3) of nottem time series.

```
nottem
  Jan Feb Mar Apr May Jun Jul Aug Sep Oct Nov Dec
1920 40.6 40.8 44.4 46.7 54.1 58.5 57.7 56.4 54.3 50.5 42.9 39.8
1921 44.2 39.8 45.1 47.0 54.1 58.7 66.3 59.9 57.0 54.2 39.7 42.8
1922 37.5 38.7 39.5 42.1 55.7 57.8 56.8 54.3 54.3 47.1 41.8 41.7
1923 41.8 40.1 42.9 45.8 49.2 52.7 64.2 59.6 54.4 49.2 36.3 37.6
1924 39.3 37.5 38.3 45.5 53.2 57.7 60.8 58.2 56.4 49.8 44.4 43.6
1925 40.0 40.5 40.8 45.1 53.8 59.4 63.5 61.0 53.0 50.0 38.1 36.3
1926 39.2 43.4 43.4 48.9 50.6 56.8 62.5 62.0 57.5 46.7 41.6 39.8
1927 39.4 38.5 45.3 47.1 51.7 55.0 60.4 60.5 54.7 50.3 42.3 35.2
1928 40.8 41.1 42.8 47.3 50.9 56.4 62.2 60.5 55.4 50.2 43.0 37.3
1929 34.8 31.3 41.0 43.9 53.1 56.9 62.5 60.3 59.8 49.2 42.9 41.9
1930 41.6 37.1 41.2 46.9 51.2 60.4 60.1 61.6 57.0 50.9 43.0 38.8
1931 37.1 38.4 38.4 46.5 53.5 58.4 60.6 58.2 53.8 46.6 45.5 40.6
1932 42.4 38.4 40.3 44.6 50.9 57.0 62.1 63.5 56.3 47.3 43.6 41.8
1933 36.2 39.3 44.5 48.7 54.2 60.8 65.5 64.9 60.1 50.2 42.1 35.8
1934 39.4 38.2 40.4 46.9 53.4 59.6 66.5 60.4 59.2 51.2 42.8 45.8
1935 40.0 42.6 43.5 47.1 50.0 60.5 64.6 64.0 56.8 48.6 44.2 36.4
1936 37.3 35.0 44.0 43.9 52.7 58.6 60.0 61.1 58.1 49.6 41.6 41.3
1937 40.8 41.0 38.4 47.4 54.1 58.6 61.4 61.8 56.3 50.9 41.4 37.1
1938 42.1 41.2 47.3 46.6 52.4 59.0 59.6 60.4 57.0 50.7 47.8 39.2
1939 39.4 40.9 42.4 47.8 52.4 58.0 60.7 61.8 58.2 46.7 46.6 37.8
```

The commands to do this are as follows with the assumption of k to be 3, 7, and 11:

```
library(forecast)
opar <- par(no.readonly=TRUE)
par(mfrow=c(2,2))
ylim <- c(min(nottem), max(nottem))
plot(nottem, main="Raw time series")
plot(ma(nottem, 3), main="Simple Moving Averages (k=3)", ylim=ylim)
plot(ma(nottem, 7), main="Simple Moving Averages (k=7)", ylim=ylim)

# plot for the 7-month moving average
plot(ma(nottem, 7), main="Simple Moving Averages (k=7)", ylim=ylim)

# plot for the 11-month moving average
plot(ma(nottem, 11), main="Simple Moving Averages (k=11)", ylim=ylim)
par(opar)
```

There is an increasing smoothness in the plot as k increases. The main concern is how to find the value of k that points out the major patterns in the data without under- or over-smoothing. This can be attained by attempting several values of k before getting the optimal one.

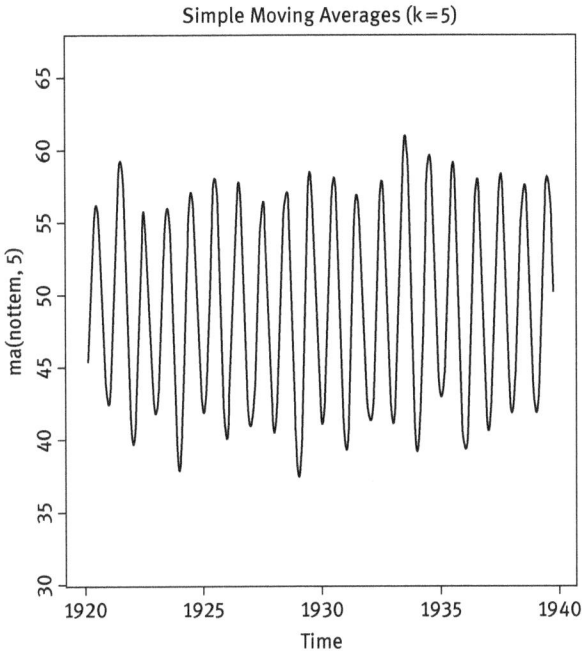

Figure 14.4: Moving average (*k*=5) of nottem time series.

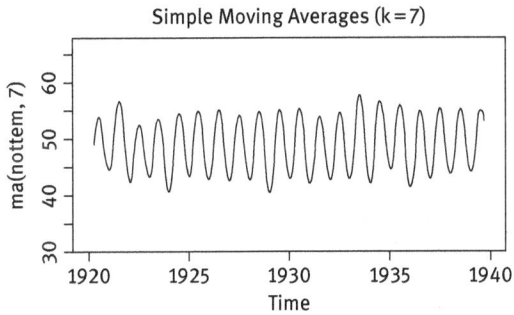

Figure 14.5: Moving average (*k* = 7) of nottem time series.

14.2.2.2 Seasonal Decomposition

Time-series data possess seasonal dimensions such as daily, weekly, monthly, or quarterly data. These data can be decomposed into a trend component, a seasonal component, and an irregular component. The trend aspect indicates changes in level over time while the seasonal components reflect cyclical

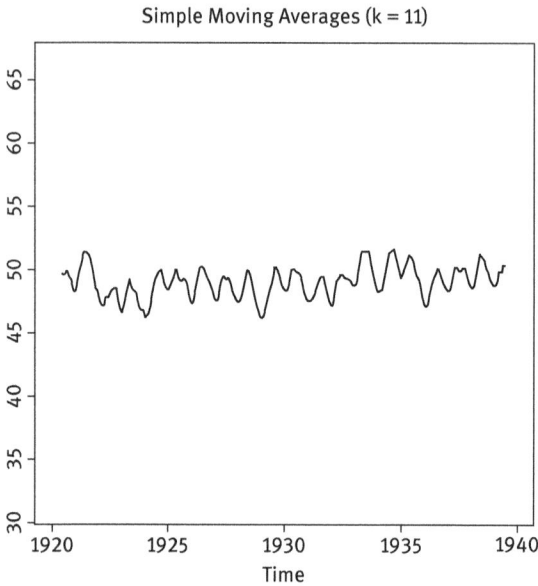

Figure 14.6: Moving average (*k*=11) of nottem time series.

effects attributed to the time of year. The irregular or error component encapsulates those impacts not captured by the trend and seasonal effects.

The decomposition can either take the form of addition or multiplication. For the case of an additive model, the components sum to provide the value of the time series. Particularly,

$$Y_t = Trend_t + Seasonal_t * Irregular_t$$

Where the observation at time t is the addition of the contributions of the trend at time t, the seasonal effect at time t, and an irregular effect at time t.

In the case of a multiplicative model, the equation is specified as:

$$Y_t = Trend_t * Seasonal_t * Irregular_t$$

The effects of each of three components are multiplied.

The main difference between additive and multiplicative models becomes more obvious through the following example. Consider a time series that records the monthly sales of rice for a period of 5 years. For the case of an additive seasonal model, the demand for bags of rice tends to rise by 300 in November and December and drop by 100 in January. The seasonal rise or fall does not depend on the current sales volume. Figure 14.7 shows the time series with different mixes of trend, seasonal and irregular components.

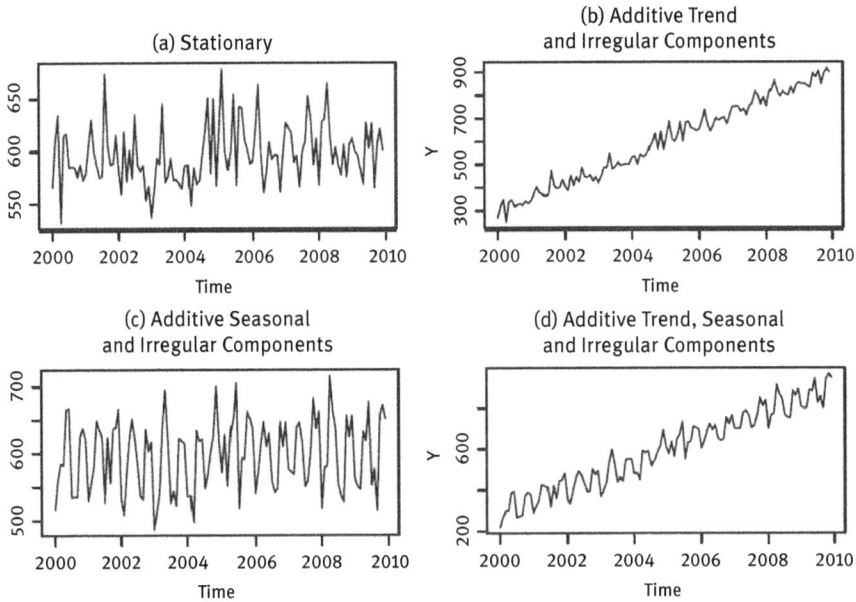

Figure 14.7: Illustrations of different time series examples with different mixes of trend, seasonal, and irregular components.

For the case of a multiplicative seasonal model, bags of rice sales in November and December tend to rise by 30 percent and decline in January by 10%. Under this model, the influence of the seasonal effect is dependent of the current sales volume. This does not occur in an additive model. Therefore, the multiplicative model is more realistic and accurate. A study of the most basic probabilities supports this conclusion. When probability of outcomes is performed additively, the tendency is to get an inaccurate conclusion.

A general approach to decompose a time series into trend, seasonal, and irregular components is seasonal decomposition by *LOESS* (locally estimated scatterplot smoothing). This can be executed in R using the *stl ( )* function. The template is:

```
stl (ts, s.window =, t.window = )
```

Where *ts* denotes the time series to be decomposed, *s.window* adjusts the speed of changes in the seasonal effects over time, and *t.window* adjusts the speed of changes in the trend over time. Smaller values encourage more rapid change. Setting *s.window* = *"periodic"* fixes seasonal effects to be the same across years. Only the *ts* and *s.window* parameters are needed.

The *stl( )* function is limited to handle additive models. However, the *stl( )* function can be applied to multiplicative models by transforming them into additive models with the use of a log transformation:

$$\log(Y_t) = \log(Trend_t * Seasonal_t * Irregular_t) = \log(Trend_t) + \log(Seasonal_t)$$
$$+ \log(Irregular_t)$$

After this transformation, the outcomes can be back-transformed to the original scale. Consider the Austres time series that records the number of Australian residents quarterly. A graph of the data is depicted in Figure 14.8. The graph shows that the fluctuation of the series rises with the level, reflecting a multiplicative model.

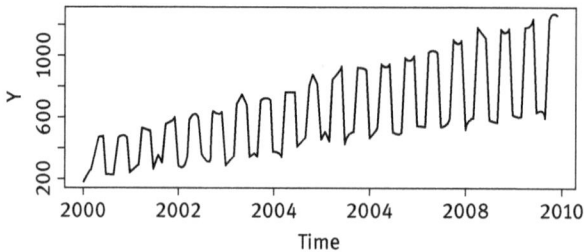

Figure 14.8: Multiplicative trend, seasonal, and irregular components chart.
Source: Excerpted from *R in Action*

The graph in Figure 14.9 reflects the time series generated by taking the log of each observation. It is characterized by stable variance, and the logged series appears to be an appropriate measure for an additive decomposition. This is executed using the following commands in R:

```
plot(Austres)
lAustres <- log(Austres)
plot(lAustres, ylab="log(Austres)")
```

The following R code is used to generate the composition graph of the Austres time series, seasonal, trend, and remainder (irregular component) shown in Figure 14.10.

```
# to plot the log of Austres time series
fit <- stl(lAustres, s.window="period")
 plot(fit)
```

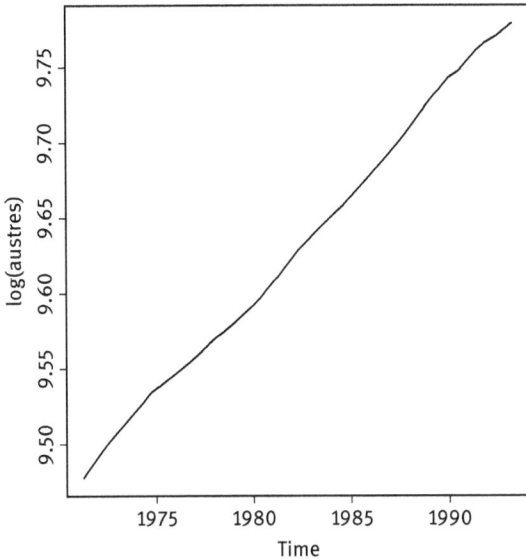

Figure 14.9: Logarithm of Austres time series.

In addition, we may be required to know the fitted values of the additive model of the Austres time series. We use *fit$time.series* because our model is stored into the variable *fit* (i.e., *fit <- stl(lAustres, s.window="period")*). Thus, the quarterly values of seasonality, trend, and remainder are displayed.

```
fit$time.series
            seasonal       trend     remainder
1971 Q2 -8.604929e-05 9.478409 -4.543850e-04
1971 Q3 -5.329708e-04 9.482872  3.535582e-04
1971 Q4  4.233129e-05 9.487312  4.965273e-04
.............................................
.............................................
.............................................
1992 Q3 -5.329708e-04 9.771603  3.707132e-04
1992 Q4  4.233129e-05 9.774125 -2.934187e-04
1993 Q1  5.766888e-04 9.776637 -2.094670e-05
1993 Q2 -8.604929e-05 9.779146  8.243940e-05
```

By logging a dataset we are converting the multiplicative relationships to additive relationships, however, we convert to linear trends by taking the exponential log of the trends. Therefore, in order to get the exponential log of the data in the Austres time series, we use the R command:

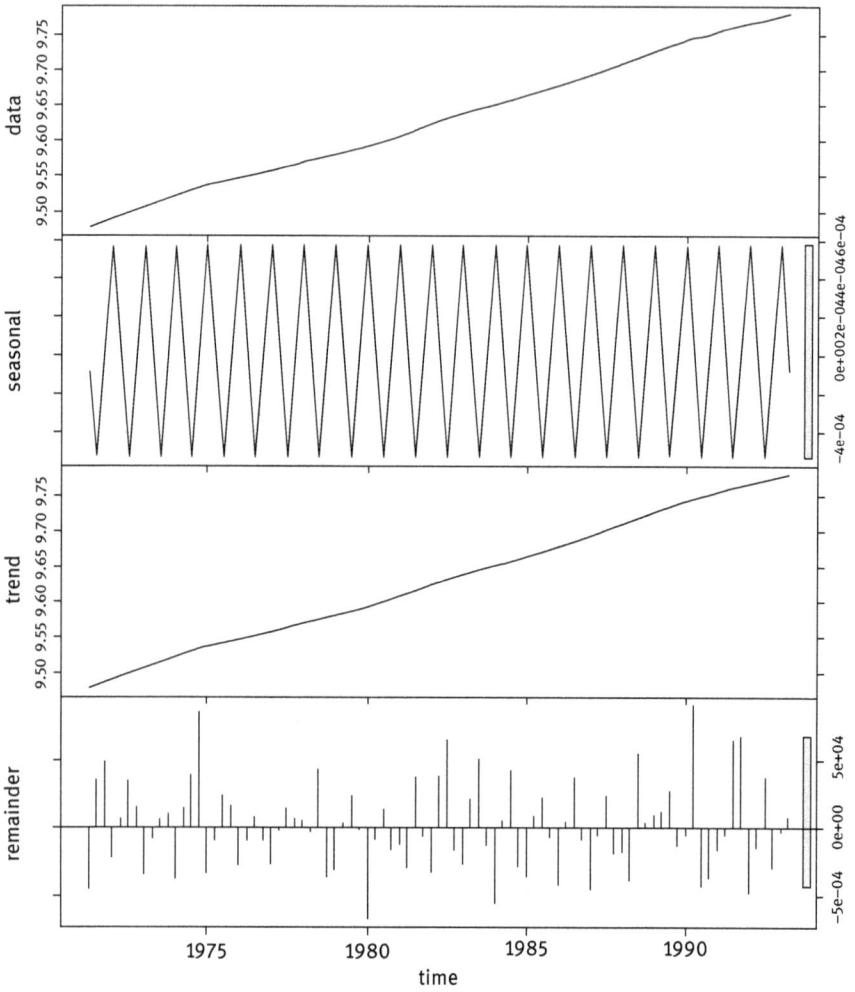

Figure 14.10: Decomposition of an Austres time series.

```
# Exponential of the log (austres time-series)
exp(fit$time.series)
 seasonal  trend   remainder
1971 Q2 0.9999140 13074.36 0.9995457
1971 Q3 0.9994672 13132.86 1.0003536
1971 Q4 1.0000423 13191.29 1.0004967
.............................................
.............................................
.............................................
```

```
1988 Q3 0.9994672 16621.36 1.0005476
1988 Q4 1.0000423 16695.63 1.0000396
1989 Q1 1.0005769 16765.87 1.0000991
```

The steps involved in decomposition of the Austres time series in Figure 14.10 are as follows and the values and exponential log of the data Austres time series are displayed above.

1. The time series is graphed and transformed
2. A seasonal decomposition is conducted and saved in an object called *fit*
3. The results are depicted in Figure 14.9
4. The *stl( )* function includes a component known as *time.series* that enables the trend, seasonal, and irregular parts of each observation
5. *Fit$time.series* depends on the logged time-series data used in the example
6. *Exp(fit$time.series)* transforms the decomposition back to the original metric

All these manipulations describes above are necessary so that we can free the data from all forms of variation, and this enables us to make a valid prediction of the dataset. All the above explanations are used to describe the time series. Any future values of the series are not yet predicted. The next section will consider how to use exponential models for forecasting beyond the available data.

14.2.2.3 Exponential Forecasting Models

Exponential models are among the general techniques used to forecast the future values of a time series. They are relatively simple and also provide good short-term predictions in a wide range of applications. However, these models vary from one another in the components of the time series that are modeled.

A *simple exponential model*, also known as a *single exponential model*, is appropriate for a time series that possesses a constant level and an irregular component at time *i*, but lacks both trend and seasonal components.

A *double exponential model*, also named a Holt exponential smoothing, is suitable for a time series that has both a level and a trend.

A *triple exponential model*, also known as a Holt-Winters exponential smoothing, is relevant for a time series that has level, trend, and seasonal components.

Exponential models can be expressed in R with either the *HoltWinters( )* function in the base installation or the *ets( )* function that comes with the forecast package. However, the *ets( )* function provides more options and is more powerful compared to the *HoltWinters( )* function.

The code for the *ets( )* function is

```
ets (ts, model = "ZZZ")
```

where *ts* is a time series and the model is expressed by three letters. The first letter captures the error type, the second letter represents the trend type, and the third letter is the seasonal type. Letter A is meant for additive, M for multiplicative, N for none, and Z for automatically selected. Table 14.3 shows functions for three exponential models. For example, *ets(ts, model = "ANN")* means that R is to use the additive error type with no trend and seasonality component, and compute estimated values using the triple exponential model. Specifically this model is referred to as simple exponential model.

Table 14.3: Functions for creating simple, double, and triple exponential forecasting models.

Type	Parameters fit	Functions
Simple	level	*ets(ts, model = "ANN")* *ses(ts)*
Double	Level, trend	*ets (ts, model = "ANN")* *holt (ts)*
Triple	Level, trend, seasonal	*ets (ts, model = " AAA")* *hw (ts)*

The *ses( )*, *holt( )*, and *hw( )* functions capture the *ets( )* function with prespecified defaults.

In order to examine these models, you need to be sure that the forecast package has been installed. If not, use this code to install it: *install.packages (" forecast")*.

14.2.2.4 Simple Exponential Smoothing

Simple exponential smoothing utilizes a weighted average (exponential moving average, or EMA) of existing time-series values to create a short-term prediction of future values. The weights are picked in order to have observations that record exponentially declining effects on the average as time adjusts backward. This is also based on the assumption that an observation in the time series can be described by:

$$X_t = level + irregular_t$$

The prediction at time X_{t+1} (called the 1-step ahead forecast) is expressed as:

$$X_{t+1} = C_0 X_t + C_1 X_{t-1} + C_2 X_{t-2} + C_3 X_{t-2} + \ldots$$

Where $c_i = \alpha (1 - \alpha)^i$, $i = 0, 1, 2, \ldots$ and $0 \leq \alpha \leq 1$. The c_i is summed to one, and the 1-step ahead forecast is defined as a weighted average of the current value and all previous values of the time series. The alpha (α) parameter controls the rate of decay for the weights. As the alpha approaches 1, more weight is assigned to recent observations. As the alpha closes to 0, more weight is assigned to past observations. The actual value of alpha is often selected by computer in order to attain an optimal fit criterion. The general fit criterion represents the sum of squared errors between the actual and predicted values.

The nottem time series records the average monthly temperature in Nottingham from 1920 to 1939 as indicated in Figure 14.1.

Owing to the absence of trend and seasonal components, the simple exponential model is appropriate to begin with. Commands for generating a 1-step ahead forecast using the *ses( )* function are as follows:

```
# Fitting of the simple exponential model
fit <- ets(nottem, model="ANN")
fit
```

The model is displayed as follows:

```
ETS(A,N,N)

Call:
 ets(y = nottem, model = "ANN")

 Smoothing parameters:
 alpha = 0.9999

 Initial states:
 l = 40.597

 sigma: 5.249

  AIC  AICc  BIC
2115.202 2115.304 2125.644
```

We may be interested in making a forecast based on the model. The R command to make a 1-step ahead forecast is given as:

```
# One-step ahead forecast from the model
forecast(fit, 1)
          Point Forecast      Lo 80     Hi 80     Lo 95    Hi 95
Jan 1940    37.80088    31.07403   44.52773  27.51305  48.08871
```

This forecast value (point forecast) is 37.80088 with lower and upper 80% CI of 31.07403 and 44.52773, respectively, while the lower and upper 95% CI is [27.51305, 48.08871].

```
# to plot the forecasted values
plot(forecast(fit,1), xlab="month", ylab=expression(paste("Temperature(",
degree*F,")"),)), main="Average Monthly Temperature")

# to print accuracy measures
accuracy(fit)

        ME    RMSE   MAE  MPE    MAPE    MASE   ACF1
Training set -0.01165165   5.227074 4.301894 -0.6269154 8.959902  1.597446 0.4518695
```

The three steps involved in the prediction are the k-step ahead forecast, the plotting of forecasted values, and checking for the accuracy of the model fitted.

The *ets(mode = "ANN")* statement links the simple exponential model to the nottem time series in Step 1. The A indicates the additive nature of the errors, and the NN denotes no trend and no seasonal component. The fit model above has value of alpha which is relatively high (0.9999) implying that distant observations are not being considered in the forecast. This value is automatically selected in order to maximize the fit of the model in relation to the given dataset.

The *forecast( )* function is explored to anticipate the time series k steps into the future with the template as *forecast(fit, k)*. The 1-step ahead forecast for this series is 37.80 with a 95% confidence interval [27.51, 48.09] as indicated in Step 2. The forecasted value of the time series is graphically depicted in Figure 14.11.

The forecast package also includes an *accuracy( )* function that shows the most common predictive accuracy measures for time-series forecasts as in Step 3. Table 14.4 defines the accuracy functions that exist in R. The e_t denotes the error or irregular component of each observation $(X_t - \hat{X}_i)$.

The *mean error* (ME) and *mean percentage error* (MPE) may not be relevant because positive errors cancel negative errors. The *root mean squared error* (RMSE) provides the square root of the mean square error which is 5.23 in this example. The *mean absolute percentage error* (MAPE) presents the error as

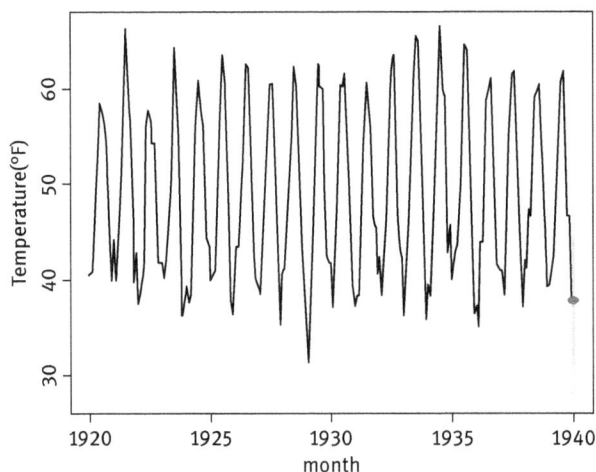

Figure 14.11: Average nottem and a 1-step ahead outlook from a simple exponential forecast with the aid of the *ets()* function.

Table 14.4: Forecasting accuracy measures.

Measure	Abbreviation	Definition
Mean error	ME	Mean (e_t)
Root mean squared error	RMSE	Sqrt $(mean(e_t^2))$
Mean absolute error	MAE	Mean $(|e_t|)$
Mean percentage error	MPE	Mean $(100 * e_t/X_t)$
Mean absolute percentage error	MAPE	Mean $(|100*e_t/X_t|)$
Mean absolute scaled error	MASE	Mean $(|w_t|)$ $w_t = \dfrac{e_t}{(\frac{1}{T-1})*sum(|X_t - X_{t-1}|)}$ T represents the number of observations, and the sum goes from t = 2 to t= T.

a percentage of the time-series values. This can be utilized to compare prediction accuracy across time series. However, it is based on the assumption of the measurement scale with a true zero point. The *mean absolute scaled error* (MASE) is the most recent accuracy indicator and is explored to compare the prediction accuracy across time series on different scales. No best measure is identified among these measures, but the *root mean squared error* (RMSE) is usually the best known and most often applied.

Note: Simple exponential smoothing is executed with the assumption of no trend and no seasonal components. The next section considers the model that realizes these assumptions.

14.2.2.5 Holt and Holt-Winters Exponential Smoothing

The *Holt exponential smoothing* technique is appropriate for a time series that has an overall level and a trend (slope). The model for an observation at time t is expressed as:

$$X_t = level + slope^*t + irregular_t$$

For this model, an alpha smoothing parameter adjusts the exponential decay for the level, and a beta smoothing parameter controls the exponential decay for the slope. The values of each parameter range between 0 and 1. The larger the value, the more weight is assigned to the most recent observations.

The *Holt-Winters exponential smoothing* technique is applicable to fit a time series that has an overall level, a trend, and a seasonal component. The model is specified as:

$$X_t = level + slope^*t + s_t + irregular_t$$

Where s_t denotes the seasonal influence at time t. The model also adds another parameter known as a *gamma smoothing* in order to control the exponential decay of the seasonal components. This parameter also ranges from 0 to 1, and as its value closes to 1, more weight is assigned to recent observations in estimating the seasonal effects.

In the preceding section, a time series of BJsales was disaggregated into additive trend, seasonal, and irregular components. Let's utilize an exponential model to forecast future inflation. In addition, the log values of the time series would be explored so that an additive model fits the data. The commands to conduct the Holt-Winters exponential smoothing technique to forecast the next five values of the BJsales time series are written:

```
#Step 1 : smoothing parameters

library(forecast)
fit <- ets(log(nottem),model="AAA")
fit

ETS(A,A,A)

Call:
 ets(y = log(nottem), model = "AAA")
```

```
Smoothing parameters:
alpha = 0.1078
beta = 1e-04
gamma = 1e-04

Initial states:
l = 3.8908
b = 0
s = -0.2024 -0.1343 0.0253   0.161   0.2274  0.2523
     0.1808  0.0852 -0.0465 -0.1406 -0.2079 -0.2004

sigma: 0.052

  AIC  AICc  BIC
-86.65098 -83.89422 -27.48012

accuracy(fit)
      ME   RMSE   MAE   MPE   MAPE  MASE  ACF1
Training set 0.0001964283 0.05020324 0.03855293 -0.009543233 1.009967 0.6607912
0.1479383

# Step 2 : Future forecasts
pred <- forecast(fit, 5)

pred
      Point Forecast Lo 80   Hi 80    Lo 95    Hi 95
Jan 1940   3.686579 3.619983 3.753175 3.584729  3.788429
Feb 1940   3.679040 3.612058 3.746023 3.576599  3.781482
Mar 1940   3.746367 3.679000 3.813735 3.643337  3.849398
Apr 1940   3.840415 3.772664 3.908167 3.736799  3.944032
May 1940   3.972161 3.904028 4.040294 3.867960  4.076362
plot(pred,main="Forecast for Temperature", ylab="Log(nottem)",xlab="Time")
```

```
pred$mean <- exp(pred$mean)
pred$lower <- exp(pred$lower)
pred$upper <- exp(pred$upper)
p <- cbind(pred$mean, pred$lower,pred$upper)
dimnames (p)[[2]] <- c("mean","Lo 80","Lo 95", "Hi 80","Hi 95")
p
         mean     Lo 80    Lo 95    Hi 80    Hi 95
Jan 1940  39.90809 37.33693 36.04359 42.65631 44.18694
Feb 1940  39.60837 37.04219 35.75175 42.35232 43.88101
Mar 1940  42.36690 39.60675 38.21917 45.31940 46.96476
Apr 1940  46.54480 43.49578 41.96343 49.80756 51.62634
May 1940  53.09915 49.60182 47.84468 56.84308 58.93070
```

Step 1 indicates the smoothing parameters for: the level (0.11), trend (0.0001), and seasonal components (0.0001). The low value for the trend does not imply no slope but shows the estimated slope from early observation was not to be updated.

The *forecast( )* function generates predictions for the next five months as indicated in Step 2, and is graphically shown in Figure 14.12. Since the forecasts are on a log scale, exponentiation is utilized to obtain the predictions in the original metric: values of the nottem as indicated in Step 3. The matrix *pred $mean* entails the point forecasts, and the matrices *pred$lower* and *pred$upper* include the 80% and 95% lower and upper confidence limits, respectively.

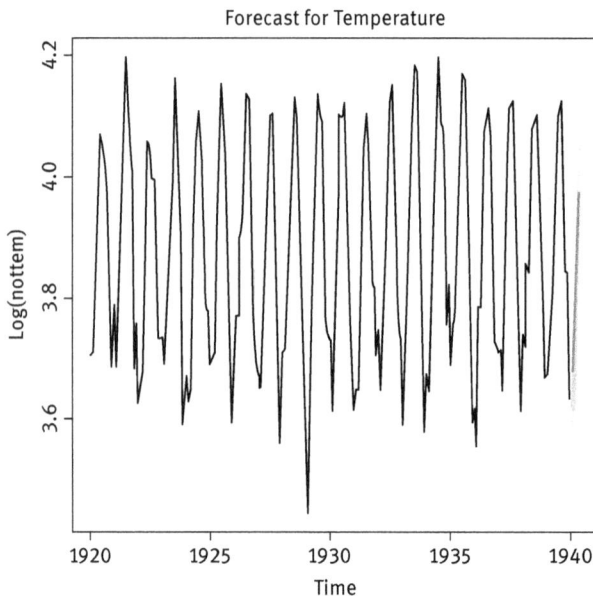

Figure 14.12: The graph of forecasted values for temperature over time.

The *exp( )* function is explored to restore the predictions to the original scale, and *cbind( )* produces a single table. Hence, the model predicts 39.61 in February 1940 with a 95% confidence band ranging from 35.75 to 43.88.

14.2.2.6 The ets( ) Function and Automated Forecasting
The *ets( )* function possesses extra capabilities in the sense that it can be explored to fit the exponential models that have multiplicative components, add a dampening component, and carrying out automated predictions. For instance,

in the preceding section, you made an additive exponential model by taking the log of the nottem time series. On the other hand, a multiplicative model can be applied to the original data. The function commands would be either *ets(nottem, model = "MAM")* or the equivalent *hw(nottem, seasonal = "multiplicative")*. Here, only the trend is considered to be additive while the seasonal and irregular components are placed under the assumption of a multiplicative nature. With the use of a multiplicative model, the accuracy statistics as well as forecast values are in the original metric (average monthly temperature)—a decided advantage.

The *ets( )* function is suitable for a dampening component. Time-series forecasts usually assume an upward trend forever (housing market, energy market). A dampening component pushes the trend to a horizontal asymptote over a period of time. In several cases, more realistic forecasts are generated from a dampened model.

All in all, the *ets( )* function can be forced to automatically choose a best-fitting model for the data. Consider an automated exponential model to the R function called JohnsonJohnson data embedded in R datasets. The selection of a best-fitting model can be obtained using the following command in R:

```
library(forecast)
fit <- ets(JohnsonJohnson)
fit
ETS(M,A,A)

Call:
 ets(y = JohnsonJohnson)

 Smoothing parameters:
 alpha = 0.2776
 beta = 0.0636
 gamma = 0.5867

 Initial states:
 l = 0.6276
 b = 0.0165
 s = -0.2293 0.1913 -0.0074 0.0454

 sigma: 0.0921

  AIC  AICc  BIC
163.6392 166.0716 185.5165

> plot(forecast(fit),main="JohnsonJohnson Forecasts", ylab="Quarterly Earnings
(Dollars)", xlab="Time", flty=2)
```

Owing to no specified model, R executes a search over a wide array of models in order to obtain one that minimizes the fit criterion (log-likelihood, by default). The chosen model is one that entails multiplicative trend, additive seasonal, and additive error components. The graph including predictions for the next quarters (the default in this case) is presented in Figure 14.13. The *flty* parameter allows the line type for the prediction line to be dashed in this example.

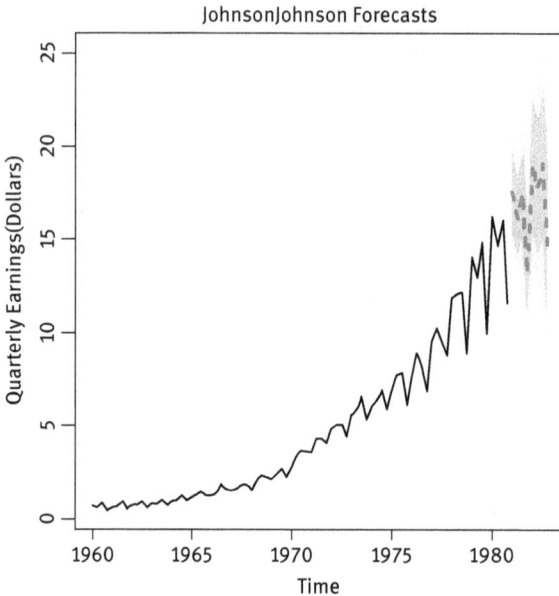

Figure 14.13: JohnsonJohnson forecast.

As mentioned at the beginning of this chapter, exponential time-series modeling is common because it can provide good short-term predictions in many situations. Another common technique is known as the Box-Jenkins methodology also referred to as ARIMA (autoregressive integrated moving average) models. ARIMA models are discussed in the next section. The purpose of these models is to determine a best fit of the time series model to the past values in the same time series.

14.2.2.7 Autoregressive Integrated Moving Average (ARIMA) Forecasting Models

In the *ARIMA forecasting model*, values are predicted linearly depend on recent values and recent errors of prediction (residuals). ARIMA is a complex technique for forecasting. This section focuses on ARIMA models for non-seasonal time series.

Terms such as lags, autocorrelation, partial correlation, differencing, and stationarity need to be explained before discussing ARIMA models.

Explanation of Concepts

Lag implies that you adjust a time series back by a given number of observations. For example, the nottem time series as well as its lags are presented in the commands below. Lag 0 is the unshifted time series. Lag 1 is the time series shifted one position to the left (down). Lag 2 shifts the time series two positions to the left (down), and so on. The lag of time series can be generated using the function *lag(ts,k)* where *ts* is the time series and *k* is the number of lags.

```
# Lag 1 of notten series
nottem_1 <- lag(nottem,1)
nottem_1
  Jan Feb Mar Apr May Jun Jul Aug Sep Oct Nov Dec
1919          40.6
1920 40.8 44.4 46.7 54.1 58.5 57.7 56.4 54.3 50.5 42.9 39.8 44.2
1921 39.8 45.1 47.0 54.1 58.7 66.3 59.9 57.0 54.2 39.7 42.8 37.5
1922 38.7 39.5 42.1 55.7 57.8 56.8 54.3 54.3 47.1 41.8 41.7 41.8
1923 40.1 42.9 45.8 49.2 52.7 64.2 59.6 54.4 49.2 36.3 37.6 39.3
1924 37.5 38.3 45.5 53.2 57.7 60.8 58.2 56.4 49.8 44.4 43.6 40.0
1925 40.5 40.8 45.1 53.8 59.4 63.5 61.0 53.0 50.0 38.1 36.3 39.2
1926 43.4 43.4 48.9 50.6 56.8 62.5 62.0 57.5 46.7 41.6 39.8 39.4
1927 38.5 45.3 47.1 51.7 55.0 60.4 60.5 54.7 50.3 42.3 35.2 40.8
1928 41.1 42.8 47.3 50.9 56.4 62.2 60.5 55.4 50.2 43.0 37.3 34.8
1929 31.3 41.0 43.9 53.1 56.9 62.5 60.3 59.8 49.2 42.9 41.9 41.6
1930 37.1 41.2 46.9 51.2 60.4 60.1 61.6 57.0 50.9 43.0 38.8 37.1
1931 38.4 38.4 46.5 53.5 58.4 60.6 58.2 53.8 46.6 45.5 40.6 42.4
1932 38.4 40.3 44.6 50.9 57.0 62.1 63.5 56.3 47.3 43.6 41.8 36.2
1933 39.3 44.5 48.7 54.2 60.8 65.5 64.9 60.1 50.2 42.1 35.8 39.4
1934 38.2 40.4 46.9 53.4 59.6 66.5 60.4 59.2 51.2 42.8 45.8 40.0
1935 42.6 43.5 47.1 50.0 60.5 64.6 64.0 56.8 48.6 44.2 36.4 37.3
1936 35.0 44.0 43.9 52.7 58.6 60.0 61.1 58.1 49.6 41.6 41.3 40.8
1937 41.0 38.4 47.4 54.1 58.6 61.4 61.8 56.3 50.9 41.4 37.1 42.1
1938 41.2 47.3 46.6 52.4 59.0 59.6 60.4 57.0 50.7 47.8 39.2 39.4
1939 40.9 42.4 47.8 52.4 58.0 60.7 61.8 58.2 46.7 46.6 37.8

# Lag 2 of notten series
nottem_2 <- lag(nottem,2)
nottem_2
  Jan Feb Mar  Apr  May Jun Jul Aug Sep Oct Nov Dec
1919          40.6 40.8
1920 44.4 46.7 54.1 58.5 57.7 56.4 54.3 50.5 42.9 39.8 44.2 39.8
1921 45.1 47.0 54.1 58.7 66.3 59.9 57.0 54.2 39.7 42.8 37.5 38.7
```

```
1922 39.5 42.1 55.7 57.8 56.8 54.3 54.3 47.1 41.8 41.7 41.8 40.1
1923 42.9 45.8 49.2 52.7 64.2 59.6 54.4 49.2 36.3 37.6 39.3 37.5
1924 38.3 45.5 53.2 57.7 60.8 58.2 56.4 49.8 44.4 43.6 40.0 40.5
1925 40.8 45.1 53.8 59.4 63.5 61.0 53.0 50.0 38.1 36.3 39.2 43.4
1926 43.4 48.9 50.6 56.8 62.5 62.0 57.5 46.7 41.6 39.8 39.4 38.5
1927 45.3 47.1 51.7 55.0 60.4 60.5 54.7 50.3 42.3 35.2 40.8 41.1
1928 42.8 47.3 50.9 56.4 62.2 60.5 55.4 50.2 43.0 37.3 34.8 31.3
1929 41.0 43.9 53.1 56.9 62.5 60.3 59.8 49.2 42.9 41.9 41.6 37.1
1930 41.2 46.9 51.2 60.4 60.1 61.6 57.0 50.9 43.0 38.8 37.1 38.4
1931 38.4 46.5 53.5 58.4 60.6 58.2 53.8 46.6 45.5 40.6 42.4 38.4
1932 40.3 44.6 50.9 57.0 62.1 63.5 56.3 47.3 43.6 41.8 36.2 39.3
1933 44.5 48.7 54.2 60.8 65.5 64.9 60.1 50.2 42.1 35.8 39.4 38.2
1934 40.4 46.9 53.4 59.6 66.5 60.4 59.2 51.2 42.8 45.8 40.0 42.6
1935 43.5 47.1 50.0 60.5 64.6 64.0 56.8 48.6 44.2 36.4 37.3 35.0
1936 44.0 43.9 52.7 58.6 60.0 61.1 58.1 49.6 41.6 41.3 40.8 41.0
1937 38.4 47.4 54.1 58.6 61.4 61.8 56.3 50.9 41.4 37.1 42.1 41.2
1938 47.3 46.6 52.4 59.0 59.6 60.4 57.0 50.7 47.8 39.2 39.4 40.9
1939 42.4 47.8 52.4 58.0 60.7 61.8 58.2 46.7 46.6 37.8

#Combination of nottem series and its lag series
nottem_group <- cbind(nottem,nottem_1,nottem_2)
nottem_group
         nottem nottem_1 nottem_2
Nov 1919   NA     NA      40.6
Dec 1919   NA     40.6    40.8
Jan 1920   40.6   40.8    44.4
.......................................
.......................................
.......................................
Oct 1939   46.7   46.6    37.8
Nov 1939   46.6   37.8    NA
Dec 1939   37.8   NA      NA
```

Autocorrelation measures the manner in which observations in a time series connect to each other. AC_k is the association between a set of observations (X_t) and observations k periods earlier (X_{t-k}). Therefore, AC_1 captures the correlation between the Lag 1 and Lag 0 time series, AC_2 represents the correlation between the Lag 2 and Lag 0 time series, and so on. The graph of these correlation $(AC_1, AC_2, \ldots, AC_k)$ is known as an *autocorrelation function* (ACF) plot. The ACF plot is utilized to choose optimal suitable parameters for the ARIMA model and also to evaluate the fit of the final model.

The ACF plot can be generated in R with the *acf( )* function in the Stats package or the *ACF( )* function in the forecast package. In this section, the *ACF( )* function is explored because it creates a more readable plot. The command is *ACF(ts)* where *ts* represents the original time series. The ACF plot for the JohnsonJohnson time series, with $k = 1$ to 4 is shown in Figure 14.16.

A partial autocorrelation is defined as the correlation between X_t and X_{t-k} with the removal of the effects of all Y values between the two $(X_{t-1}, X_{t-2}, \ldots, X_{t-k+1})$. Partial autocorrelations can also be graphed for multiple values of k. The partial autocorrelation function (PACF) plot (partial correlation coefficients between the series and lags of the series) can be produced with either the *pacf( )* function in the stats package or the *Pacf( )* function in the forecast package. In addition, the *Pacf( )* function is used because of its formatting. The function command is *Pacf(ts)*, where *ts* denotes the concerned time series. The PACF plot is also utilized to obtain the most appropriate parametes for the ARIMA model. The outcomes for the JohnsonJohnson time series are displayed in Figure 14.17.

ARIMA models are built to fit stationary time series or time series that can be transformed into stationary series. In the case of a stationary time series, the statistical features of the series do not vary over time. For instance, the mean and variance of X_t remain unchanged. Furthermore, the autocorrelations for any lag k remains constant with time.

Prior to designing an ARIMA model, it is very important to transform the values of a time series to obtain constant variance. The log transformation is usually used to achieve this as explained in section 14.2.2. The Box Cox transformation is another technique to get constant variance.

Since stationary time series is based on the assumption of constant means, it is difficult for them to have a trend component. The transformation of many non-stationary time series to stationary is conducted through *differencing*. In differencing, each value of a time series X_t is substituted with $X_{t-1} - X_t$. This helps remove a linear trend. By taking the second differencing, a quadratic trend is removed. A third differencing eliminates a cubic trend. The differencing does not often exceed the second time.

The *diff( )* function is used to difference a time series, with the format as *diff(ts, differences =d)* where d is the number of times the time series *ts* has to be differenced. The default is $d = 1$. The best value of d can be determined by the use of the *ndiffs( )* function in the forecast package.

Investigation of stationarity is usually based on a visual inspection of a time-series plot. In the absence of constant variance, the data are transformed. The data is differenced if trends appear. The statistical step known as the Augmented Dickey Fuller (ADF) test (a test of the null hypothesis, suggesting that the greater the negative result, the stronger the likely rejection) is utilized to assess the assumption of stationarity. The ADF test is performed in R with the *adf.test( )* function in the tseries package. The command is *adf.test (ts)*, where *ts* is the times series to be examined. A significant result indicates stationarity.

In summary, ACF and PCF plots are explored to obtain the parameters of ARIMA models. Stationarity is a fundamental assumption while transformation and differencing are utilized as helping tools to achieve stationarity. Having explained these concepts, the next section discusses the design of models with an autoregressive (AR) component, a moving averages (MA) component, or both components (autoregressive moving average, or ARMA). In the concluding part of this section, ARIMA models that entail ARMA component and differencing to obtain stationarity (Integration) will be examined.

14.3 Multivariate Time Series Models

A multivariate time series model has more than one time-dependent variable and each of the variables depends on its past values and also depends on other variables. It is used to model and explain the interactions among a group of time series variables, and it is used to predict the future values. For example, customer knowledge, purchase intention, celebrity endorsement and perceived value are the factors that influenced customer's purchase intention of a product. We can say that these multiple variables are to be considered in predicting customer's purchase intention.

14.3.1 ARMA and ARIMA Models

For the case of an autoregressive model of order p, each value in a time series is forecasted from a linear combination of the past p values:

$$AR(p): X_t = \mu + \beta_1 X_{t-1} + \beta_2 X_{t-2} + \ldots + \beta_p X_{t-p} + \varepsilon_t$$

Where X_t denotes a given value of the series, μ is the mean of the series, the β_s represent the weights, and ε_t is the irregular component. For a moving average model of order q, each value in the time series is anticipated from a linear combination of q previous errors. This is specified as:

$$MA(q): X_t = \mu - \gamma_1 \varepsilon_{t-1} - \gamma_2 \varepsilon_{t-2} - \ldots - \gamma_p \varepsilon_{t-q} + \varepsilon_t$$

Where the ε_s represent the errors of prediction and the γ_s are the weights. It is important to note that the moving averages here are not the same as the simple moving averages discussed in Section 14.2. In this application, the MS models univariate time series and depends linearly on current and past values, rather than a simple averaging effect.

Integrating these models give birth to an *ARMA(p,q)* model which is mathematically defined as:

$$X_t = \mu + \beta_1 X_{t-1} + \beta_2 X_{t-2} + \ldots + \beta_p X_{t-p} - \gamma_1 \varepsilon_{t-1} - \gamma_2 \varepsilon_{t-2} - \ldots - \gamma_p \varepsilon_{t-q} + \varepsilon_t$$

The ARMA model forecasts each value of the time series from the past p values and q residuals.

An *ARIMA(p, d,q)* model is regarded as a model in which time series have been differenced d times, and the resulting values are forecasted from the past p actual values and q previous errors. The forecasts are "un-differenced," or integrated to obtain the final prediction.

The following steps are involved in ARIMA modelling:
1. Ensure that the time series is stationary
2. Find an appropriate model or models with possible values of p and q
3. Design the model
4. Assess the model's fit based on statistical assumptions and predictive accuracy
5. Make predictions

Let's apply these steps in relation to the BJsales time series in order to fit an ARIMA model.

Step 1: Ensuring that the time series is stationary
You need to first plot the time series and evaluate its stationarity (the plot is shown in Figure 14.14). The variance seems to be constant across the years observed, thus eliminating the need for a transformation. However, there is likely to be a trend, buttressed by the outcomes of the *ndiffs( )* function.

```
# to get the maximum difference of the time series JohnsonJohnson dataset before
it can be stationary
ndiffs(JohnsonJohnson)
[1] 1
# since the series is stationary at the first difference, we store it into
dJohnsonJohnson and then plot the graph
dJohnsonJohnson <- diff(JohnsonJohnson)
plot(dJohnsonJohnson)
```

```
# test for stationarity of the first difference of JohnsonJohnson using Augented
Dickey Fuller test
adf.test(dJohnsonJohnson)
```

```
 Augmented Dickey-Fuller Test
data: dJohnsonJohnson
Dickey-Fuller = -3.9886, Lag order = 4, p-value = 0.01421
alternative hypothesis: stationary
```

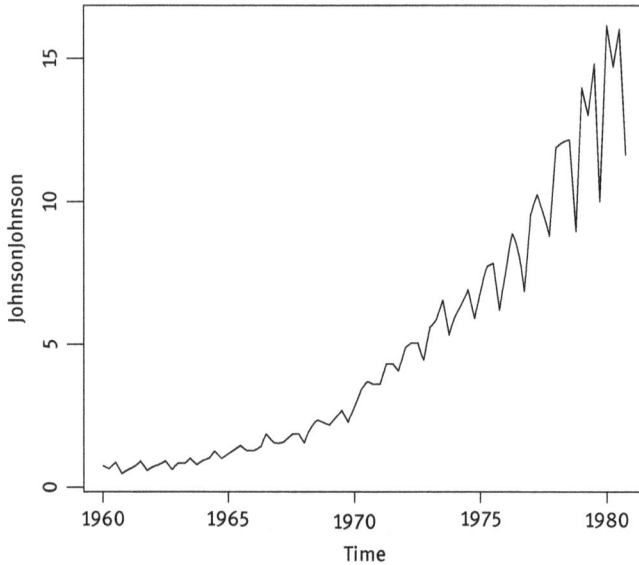

Figure 14.14: JohnsonJohnson time series.
plot(JohnsonJohnson)

A first difference is taken for the series (*lag=1* is the default) and saved as dJohnsonJohnson. The differenced time series is graphed in Figure 14.15, and seems to be more stationary. With the use of the ADF test to the differenced series, the outcome reveals the presence of stationarity.

Step 2: Finding one or more appropriate models
Reasonable models are chosen in line with the ACF and PACF plots. The commands used for these plots are as follows:

acf(dJohnsonJohnson)
pacf(dJohnsonJohnson)

The resulting plots are depicted in Figures 14.16 and 14.17.

The aim is to determine the values for the parameters *p,d,q*. Since the value for *d* is already known to be 1, the values for *p* and *q* can be obtained by comparing the ACF and PACF plots based on the rules in Table 14.5.

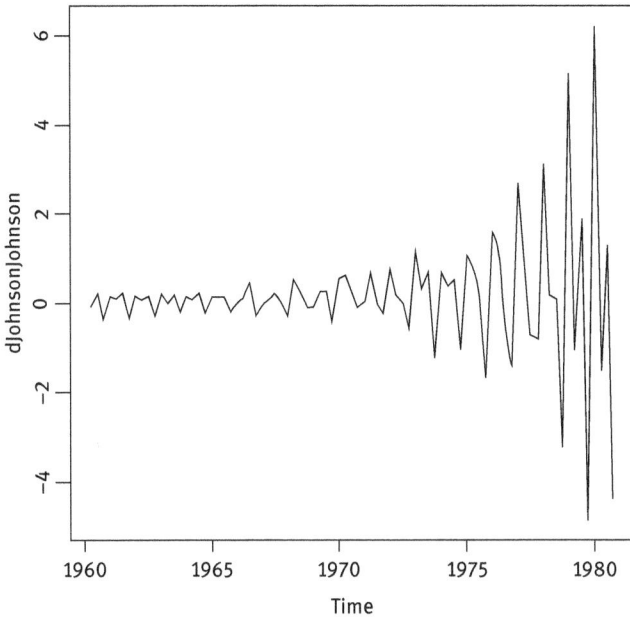

Figure 14.15: First difference of JohnsonJohnson time series.

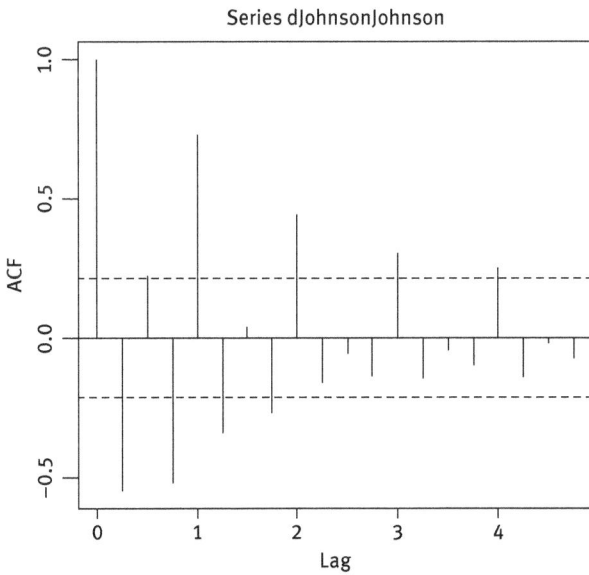

Figure 14.16: Autocorrelation Function (ACF) of first difference of JohnsonJohnson time series.

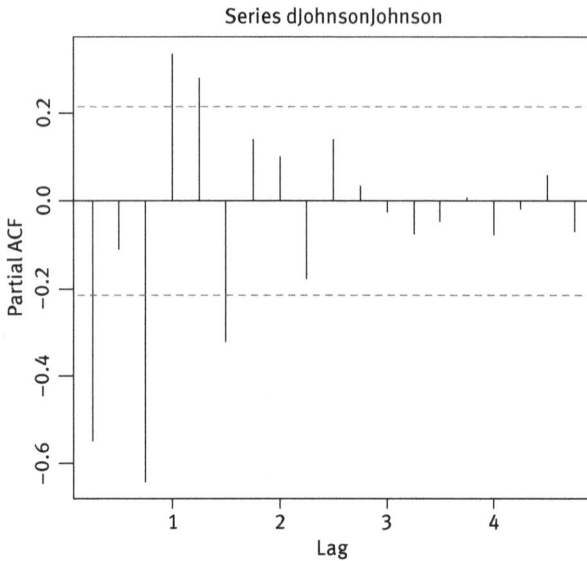

Figure 14.17: Partial Autocorrelation Function (PACF) of first difference of JohnsonJohnson time series.

Table 14.5: Rules for choosing an ARIMA model.

Model	ACF	PACF
ARIMA (p, d, 0)	Paths off to zero	Zero after lag p
ARIMA(0,d,q)	Zero after lag q	Paths off to zero
ARIMA(p, d, q)	Paths off to zero	Paths off to zero

The results in Table 14.5 seem to be more theoretical, implying the actual ACF and PACF may not follow the guidelines. However, it can be used to a rough guide in attempting to find reasonable models. For the JohnsonJohnson time series in Figures 14.16 and 14.17, there seems to be one large autocorrelation at lag 1, and the partial autocorrelations path off to zero as the lags get bigger. This suggests an attempt to use an ARIMA(0,1,1) model.

Step 3: Designing the model
The ARIMA model is designed with the *arima( )* function. The code is *arima(ts, order = c(p,d,q)*. The finding of designing an ARIMA(0,1,1) model to the JohnsonJohnson time series is:

```
library(forecast)
fit <- arima(JohnsonJohnson, order=c(0,1,1))

# to display the model
fit

Call:
arima(x = JohnsonJohnson, order = c(0, 1, 1))

Coefficients:
   ma1
 -0.5732
s.e.  0.0634

sigma^2 estimated as 1.36: log likelihood = -130.74, aic = 265.48

# to compute the accuracy measures
 accuracy(fit)
     ME    RMSE   MAE    MPE   MAPE   MASE   ACF1
Training set 0.3577933 1.159352 0.7281882  5.7376  14.62167 0.9332863  -0.3019172
```

It is critical to remember that the model is applied to the original time series. With the specification of *d*=1, R generates the first difference for you. The coefficient for the moving average is -0.57 which is reported along with the *Akaike information criterion* (AIC), a measure of the relative quality of a model for a given data set.. The AIC can assist in selecting which model is most reasonable in the midst of other models. The smaller the value of the AIC, the better the model is.

The accuracy measures would help us determine whether the model is appropriate with sufficient accuracy. Here, the mean absolute percent error is about 15% of quarterly earnings.

Step 4: Assessing the model fit
An appropriate model is expected to have residuals that are normally distributed with mean zero, and zero autocorrelation for every possible lag. Put differently, the residuals need to be normally and independently distributed with no relationship between them. The following commands are used to assess these assumptions:

```
# Assessing the model fit
qqnorm(fit$residuals)
qqline(fit$residuals)
Box.test(fit$residuals, type ="Ljung-Box")
```

```
Box-Ljung test
data: fit$residuals
X-squared = 7.9337, df = 1, p-value = 0.004852
```

The *qqnorm( )* and *qqline( )* functions generate the plot in Figure 14.18. Normally distributed data have to fall along the line. In this example, the results appear good.

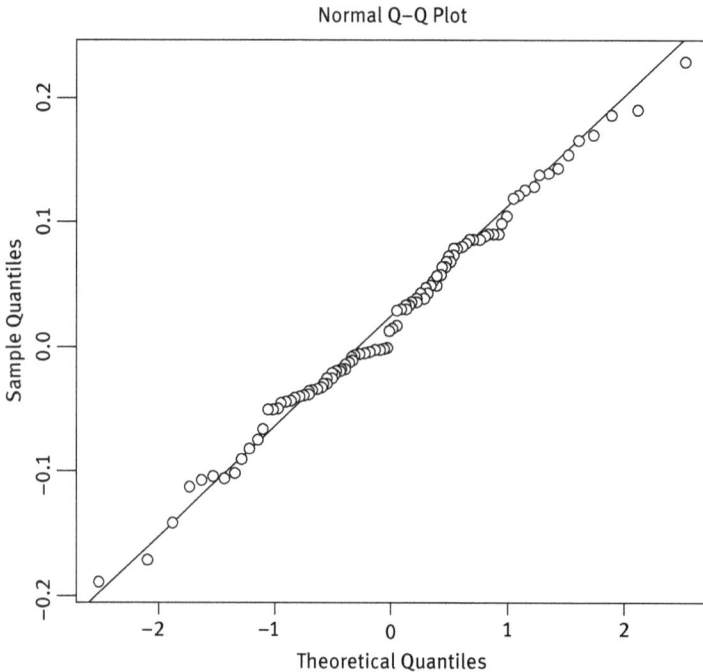

Figure 14.18: QQnorm and QQline functions of first difference of JohnsonJohnson time series.

The *Box.test( )* function presents a test that the autocorrelations are all zero. The results are insignificant, implying that the autocorrelations are not different from zero. This ARIMA model seems to fit the data well.

Step 5: Making predictions
If the model fails to satisfy the assumptions of normal residuals and zero auto-correlation, this calls for a need to change the model, add parameters, or try a different method. Having selected the final model, it is more important to make forecasts of future values. In the next procedure, the *forecast( )* function in the forecast package is employed to forecast three years ahead.

```
forecast(fit,3)
  Point Forecast Lo 80 Hi 80 Lo 95 Hi 95
1981 Q1   13.52533 12.03063 15.02002 11.23939 15.81126
1981 Q2   13.52533 11.90021 15.15044 11.03993 16.01072
1981 Q3   13.52533 11.77951 15.27114 10.85533 16.19532
plot(forecast(fit,3), xlab="Quarter", ylab="Earnings")
```

The *plot( )* function is used to graph the prediction in Figure 14.18. Point estimates are reported with the blue dots while 80% and 90% confidence bands are indicated by dark and light bands, respectively.

14.3.2 Automated ARIMA Predictions

In the preceding section, the *ets( )* function in the forecast package is explored to automatically choose the best exponential model. The package also includes an *auto.arima( )* function to choose the best ARIMA model. This section applies this technique.

```
library(forecast)
fit <- auto.arima(JohnsonJohnson)
fit
Series: JohnsonJohnson
ARIMA(1,1,2)(0,1,0)[4]

Coefficients:
   ar1   ma1   ma2
  -0.7921 -0.0970 -0.3945
s.e. 0.1396 0.1802 0.1580

sigma^2 estimated as 0.1834: log likelihood=-44.07
AIC=96.14 AICc=96.68 BIC=105.61

forecast(fit, 3)
  Point Forecast Lo 80 Hi 80 Lo 95 Hi 95
1981 Q1   17.78046 17.23158 18.32934 16.94102 18.61990
1981 Q2   16.25397 15.70173 16.80621 15.40939 17.09856
1981 Q3   17.60119 17.00261 18.19976 16.68575 18.51663

accuracy(fit)
        ME     RMSE     MAE    MPE  MAPE   MASE     ACF1
Training set 0.06773422 0.4073877 0.2676043 2.0526 6.500665 0.3814065 0.01008865
```

The function picks an ARIMA model with $p=2$, $d=1$, and $q=2$. These are values that minimize the AIC criterion over a large number of possible models. The MPE and MAPE accuracy increase due to the presence of zero values in the series (a disadvantage of these two statistics).

Prediction has a long and varied history, from early shamans forecasting the weather to modern data scientists that engage in predicting the stock markets. Prediction is attached to both science and human nature. This chapter examined how to create time series in R, evaluate trends, and investigate seasonal effects. The two most common techniques, exponential models and ARIMA models, were explored in making predictions.

These techniques are relevant in understanding and predicting a wide variety of phenomena. However, it is more important to know that each includes extrapolation—going beyond the data. Their assumption is that future conditions reflect current conditions. For instance, financial forecasts established in 2007 with the assumption of continued economic growth in 2008 and beyond failed because significant events altered the trend and pattern in a time series. The farther out you attempt to predict, the greater the uncertainty.

Exercises for Chapter 14

1. Discuss two main approaches in time series analysis.
2. What is the relevance of time series analysis?
3. Create a time series object based on the average monthly crude oil price between January 2015 and December 2017.
4. Plot the time series object from Exercise 3.
5. Create a subset time series of Exercise 3 that starts from May 2015 and ends May 2017.
6. Find an appropriate ARIMA model using the R inbuilt dataset of AirPassengers.

References

Kabacoff, R. (2015). *R in action: Data analysis and graphics with* R. Shelter Island, NY: Manning.
Younus, S., Rasheed, F., and Zia, A. (2015). Identifying the Factors Affecting Customer
    Purchase Intention. *Global Journal of Management and Business Research*, 15(2), 8–14.

Websites:

https://www.emathzone.com
https://www.intmath.com/counting-probability/13-poisson-probability-distribution.php#mean
https://www.umass.edu/wsp/resources/poisson/#end
https://businessjargons.com/chi-square-distribution.html

https://doi.org/10.1515/9781547401475-015

APPENDIX

Table of standard normal distribution values ($z \le 0$)

$-\zeta$	0	0.01	0.02	0.03	0.04	0.05	0.06	0.07	0.08	0.09
0	0.5	0.49601	0.49202	0.48803	0.48405	0.48006	0.47608	0.4721	0.46812	0.46414
0.1	0.46017	0.45621	0.45224	0.44828	0.44433	0.44038	0.43644	0.43251	0.42858	0.42466
0.2	0.42074	0.41683	0.41294	0.40905	0.40517	0.40129	0.39743	0.39358	0.38974	0.38591
0.3	0.38209	0.37828	0.37448	0.3707	0.36693	0.36317	0.35942	0.35569	0.35197	0.34827
0.4	0.34458	0.3409	0.33724	0.3336	0.32997	0.32636	0.32276	0.31918	0.31561	0.31207
0.5	0.30854	0.30503	0.30153	0.29806	0.2946	0.29116	0.28774	0.28434	0.28096	0.2776
0.6	0.27425	0.27093	0.26763	0.26435	0.26109	0.25785	0.25463	0.25143	0.24825	0.2451
0.7	0.24196	0.23885	0.23576	0.2327	0.22965	0.22663	0.22363	0.22065	0.2177	0.21476
0.8	0.21186	0.20897	0.20611	0.20327	0.20045	0.19766	0.19489	0.19215	0.18943	0.18673
0.9	0.18406	0.18141	0.17879	0.17619	0.17361	0.17106	0.16853	0.16602	0.16354	0.16109
1	0.15866	0.15625	0.15386	0.15151	0.14917	0.14686	0.14457	0.14231	0.14007	0.13786
1.1	0.13567	0.1335	0.13136	0.12924	0.12714	0.12507	0.12302	0.121	0.119	0.11702
1.2	0.11507	0.11314	0.11123	0.10935	0.10749	0.10565	0.10384	0.10204	0.10027	0.09853
1.3	0.0968	0.0951	0.09342	0.09176	0.09012	0.08851	0.08692	0.08534	0.08379	0.08226
1.4	0.08076	0.07927	0.0778	0.07636	0.07493	0.07353	0.07215	0.07078	0.06944	0.06811
1.5	0.06681	0.06552	0.06426	0.06301	0.06178	0.06057	0.05938	0.05821	0.05705	0.05592
1.6	0.0548	0.0537	0.05262	0.05155	0.0505	0.04947	0.04846	0.04746	0.04648	0.04551
1.7	0.04457	0.04363	0.04272	0.04182	0.04093	0.04006	0.0392	0.03836	0.03754	0.03673
1.8	0.03593	0.03515	0.03438	0.03363	0.03288	0.03216	0.03144	0.03074	0.03005	0.02938
1.9	0.02872	0.02807	0.02743	0.0268	0.02619	0.02559	0.025	0.02442	0.02385	0.0233
2	0.02275	0.02222	0.02169	0.02118	0.02068	0.02018	0.0197	0.01923	0.01876	0.01831
2.1	0.01786	0.01743	0.017	0.01659	0.01618	0.01578	0.01539	0.015	0.01463	0.01426
2.2	0.0139	0.01355	0.01321	0.01287	0.01255	0.01222	0.01191	0.0116	0.0113	0.01101
2.3	0.01072	0.01044	0.01017	0.0099	0.00964	0.00939	0.00914	0.00889	0.00866	0.00842
2.4	0.0082	0.00798	0.00776	0.00755	0.00734	0.00714	0.00695	0.00676	0.00657	0.00639
2.5	0.00621	0.00604	0.00587	0.0057	0.00554	0.00539	0.00523	0.00509	0.00494	0.0048
2.6	0.00466	0.00453	0.0044	0.00427	0.00415	0.00403	0.00391	0.00379	0.00368	0.00357
2.7	0.00347	0.00336	0.00326	0.00317	0.00307	0.00298	0.00289	0.0028	0.00272	0.00264
2.8	0.00256	0.00248	0.0024	0.00233	0.00226	0.00219	0.00212	0.00205	0.00199	0.00193
2.9	0.00187	0.00181	0.00175	0.0017	0.00164	0.00159	0.00154	0.00149	0.00144	0.0014

https://doi.org/10.1515/9781547401475-016

(continued)

-ζ	0	0.01	0.02	0.03	0.04	0.05	0.06	0.07	0.08	0.09
3	0.00135	0.00131	0.00126	0.00122	0.00118	0.00114	0.00111	0.00107	0.00104	0.001
3.1	0.00097	0.00094	0.0009	0.00087	0.00085	0.00082	0.00079	0.00076	0.00074	0.00071
3.2	0.00069	0.00066	0.00064	0.00062	0.0006	0.00058	0.00056	0.00054	0.00052	0.0005
3.3	0.00048	0.00047	0.00045	0.00043	0.00042	0.0004	0.00039	0.00038	0.00036	0.00035
3.4	0.00034	0.00033	0.00031	0.0003	0.00029	0.00028	0.00027	0.00026	0.00025	0.00024
3.5	0.00023	0.00022	0.00022	0.00021	0.0002	0.00019	0.00019	0.00018	0.00017	0.00017

Table of standard normal distribution values ($z \geq 0$)

ζ	0	0.01	0.02	0.03	0.04	0.05	0.06	0.07	0.08	0.09
0	0.5	0.50399	0.50798	0.51197	0.51595	0.51994	0.52392	0.5279	0.53188	0.53586
0.1	0.53983	0.5438	0.54776	0.55172	0.55567	0.55962	0.56356	0.56749	0.57142	0.57535
0.2	0.57926	0.58317	0.58706	0.59095	0.59483	0.59871	0.60257	0.60642	0.61026	0.61409
0.3	0.61791	0.62172	0.62552	0.6293	0.63307	0.63683	0.64058	0.64431	0.64803	0.65173
0.4	0.65542	0.6591	0.66276	0.6664	0.67003	0.67364	0.67724	0.68082	0.68439	0.68793
0.5	0.69146	0.69497	0.69847	0.70194	0.7054	0.70884	0.71226	0.71566	0.71904	0.7224
0.6	0.72575	0.72907	0.73237	0.73565	0.73891	0.74215	0.74537	0.74857	0.75175	0.7549
0.7	0.75804	0.76115	0.76424	0.7673	0.77035	0.77337	0.77637	0.77935	0.7823	0.78524
0.8	0.78814	0.79103	0.79389	0.79673	0.79955	0.80234	0.80511	0.80785	0.81057	0.81327
0.9	0.81594	0.81859	0.82121	0.82381	0.82639	0.82894	0.83147	0.83398	0.83646	0.83891
1	0.84134	0.84375	0.84614	0.84849	0.85083	0.85314	0.85543	0.85769	0.85993	0.86214
1.1	0.86433	0.8665	0.86864	0.87076	0.87286	0.87493	0.87698	0.879	0.881	0.88298
1.2	0.88493	0.88686	0.88877	0.89065	0.89251	0.89435	0.89617	0.89796	0.89973	0.90147
1.3	0.9032	0.9049	0.90658	0.90824	0.90988	0.91149	0.91308	0.91466	0.91621	0.91774
1.4	0.91924	0.92073	0.9222	0.92364	0.92507	0.92647	0.92785	0.92922	0.93056	0.93189
1.5	0.93319	0.93448	0.93574	0.93699	0.93822	0.93943	0.94062	0.94179	0.94295	0.94408
1.6	0.9452	0.9463	0.94738	0.94845	0.9495	0.95053	0.95154	0.95254	0.95352	0.95449
1.7	0.95543	0.95637	0.95728	0.95818	0.95907	0.95994	0.9608	0.96164	0.96246	0.96327
1.8	0.96407	0.96485	0.96562	0.96638	0.96712	0.96784	0.96856	0.96926	0.96995	0.97062
1.9	0.97128	0.97193	0.97257	0.9732	0.97381	0.97441	0.975	0.97558	0.97615	0.9767

(continued)

ζ	0	0.01	0.02	0.03	0.04	0.05	0.06	0.07	0.08	0.09
2	0.97725	0.97778	0.97831	0.97882	0.97932	0.97982	0.9803	0.98077	0.98124	0.98169
2.1	0.98214	0.98257	0.983	0.98341	0.98382	0.98422	0.98461	0.985	0.98537	0.98574
2.2	0.9861	0.98645	0.98679	0.98713	0.98745	0.98778	0.98809	0.9884	0.9887	0.98899
2.3	0.98928	0.98956	0.98983	0.9901	0.99036	0.99061	0.99086	0.99111	0.99134	0.99158
2.4	0.9918	0.99202	0.99224	0.99245	0.99266	0.99286	0.99305	0.99324	0.99343	0.99361
2.5	0.99379	0.99396	0.99413	0.9943	0.99446	0.99461	0.99477	0.99492	0.99506	0.9952
2.6	0.99534	0.99547	0.9956	0.99573	0.99585	0.99598	0.99609	0.99621	0.99632	0.99643
2.7	0.99653	0.99664	0.99674	0.99683	0.99693	0.99702	0.99711	0.9972	0.99728	0.99736
2.8	0.99744	0.99752	0.9976	0.99767	0.99774	0.99781	0.99788	0.99795	0.99801	0.99807
2.9	0.99813	0.99819	0.99825	0.99831	0.99836	0.99841	0.99846	0.99851	0.99856	0.99861
3	0.99865	0.99869	0.99874	0.99878	0.99882	0.99886	0.99889	0.99893	0.99896	0.999
3.1	0.99903	0.99906	0.9991	0.99913	0.99916	0.99918	0.99921	0.99924	0.99926	0.99929
3.2	0.99931	0.99934	0.99936	0.99938	0.9994	0.99942	0.99944	0.99946	0.99948	0.9995
3.3	0.99952	0.99953	0.99955	0.99957	0.99958	0.9996	0.99961	0.99962	0.99964	0.99965
3.4	0.99966	0.99968	0.99969	0.9997	0.99971	0.99972	0.99973	0.99974	0.99975	0.99976
3.5	0.99977	0.99978	0.99978	0.99979	0.9998	0.99981	0.99981	0.99982	0.99983	0.99983

Table of the Student's *t*-distribution

The table gives the values of $t_{\alpha;\nu}$ where
$\Pr(T_\nu > t_{\alpha;\nu}) = \alpha$, with ν degrees of freedom

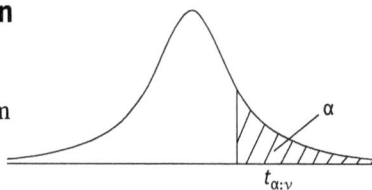

α / ν	0.1	0.05	0.025	0.01	0.005	0.001	0.0005
1	3.078	6.314	12.076	31.821	63.657	318.310	636.620
2	1.886	2.920	4.303	6.965	9.925	22.326	31.598
3	1.638	2.353	3.182	4.541	5.841	10.213	12.924
4	1.533	2.132	2.776	3.747	4.604	7.173	8.610
5	1.476	2.015	2.571	3.365	4.032	5.893	6.869
6	1.440	1.943	2.447	3.143	3.707	5.208	5.959

(continued)

v \ α	0.1	0.05	0.025	0.01	0.005	0.001	0.0005
7	1.415	1.895	2.365	2.998	3.499	4.785	5.408
8	1.397	1.860	2.306	2.896	3.355	4.501	5.041
9	1.383	1.833	2.262	2.821	3.250	4.297	4.781
10	1.372	1.812	2.228	2.764	3.169	4.144	4.587
11	1.363	1.796	2.201	2.718	3.106	4.025	4.437
12	1.356	1.782	2.179	2.681	3.055	3.930	4.318
13	1.350	1.771	2.160	2.650	3.012	3.852	4.221
14	1.345	1.761	2.145	2.624	2.977	3.787	4.140
15	1.341	1.753	2.131	2.602	2.947	3.733	4.073
16	1.337	1.746	2.120	2.583	2.921	3.686	4.015
17	1.333	1.740	2.110	2.567	2.898	3.646	3.965
18	1.330	1.734	2.101	2.552	2.878	3.610	3.922
19	1.328	1.729	2.093	2.539	2.861	3.579	3.883
20	1.325	1.725	2.086	2.528	2.845	3.552	3.850
21	1.323	1.721	2.080	2.518	2.831	3.527	3.819
22	1.321	1.717	2.074	2.508	2.819	3.505	3.792
23	1.319	1.714	2.069	2.500	2.807	3.485	3.767
24	1.318	1.711	2.064	2.492	2.797	3.467	3.745
25	1.316	1.708	2.060	2.485	2.787	3.450	3.725
26	1.315	1.706	2.056	2.479	2.779	3.435	3.707
27	1.314	1.703	2.052	2.473	2.771	3.421	3.690
28	1.313	1.701	2.048	2.467	2.763	3.408	3.674
29	1.311	1.699	2.045	2.462	2.756	3.396	3.659
30	1.310	1.697	2.042	2.457	2.750	3.385	3.646
40	1.303	1.684	2.021	2.423	2.704	3.307	3.551
60	1.296	1.671	2.000	2.390	2.660	3.232	3.460
120	1.289	1.658	1.980	2.358	2.617	3.160	3.373
∞	1.282	1.645	1.960	2.326	2.576	3.090	3.291

F-distribution (Upper tail probability = 0.05) Numerator df = 1 to 10

df2\df1	1	2	3	4	5	6	7	8	10
1	161.448	199.500	215.707	224.583	230.162	233.986	236.768	238.883	241.882
2	18.513	19.000	19.164	19.247	19.296	19.330	19.353	19.371	19.396
3	10.128	9.552	9.277	9.117	9.013	8.941	8.887	8.845	8.786
4	7.709	6.944	6.591	6.388	6.256	6.163	6.094	6.041	5.964
5	6.608	5.786	5.409	5.192	5.050	4.950	4.876	4.818	4.735
6	5.987	5.143	4.757	4.534	4.387	4.284	4.207	4.147	4.060
7	5.591	4.737	4.347	4.120	3.972	3.866	3.787	3.726	3.637
8	5.318	4.459	4.066	3.838	3.687	3.581	3.500	3.438	3.347
9	5.117	4.256	3.863	3.633	3.482	3.374	3.293	3.230	3.137
10	4.965	4.103	3.708	3.478	3.326	3.217	3.135	3.072	2.978
11	4.844	3.982	3.587	3.357	3.204	3.095	3.012	2.948	2.854
12	4.747	3.885	3.490	3.259	3.106	2.996	2.913	2.849	2.753
13	4.667	3.806	3.411	3.179	3.025	2.915	2.832	2.767	2.671
14	4.600	3.739	3.344	3.112	2.958	2.848	2.764	2.699	2.602
15	4.543	3.682	3.287	3.056	2.901	2.790	2.707	2.641	2.544
16	4.494	3.634	3.239	3.007	2.852	2.741	2.657	2.591	2.494
17	4.451	3.592	3.197	2.965	2.810	2.699	2.614	2.548	2.450
18	4.414	3.555	3.160	2.928	2.773	2.661	2.577	2.510	2.412
19	4.381	3.522	3.127	2.895	2.740	2.628	2.544	2.477	2.378
20	4.351	3.493	3.098	2.866	2.711	2.599	2.514	2.447	2.348
21	4.325	3.467	3.072	2.840	2.685	2.573	2.488	2.420	2.321
22	4.301	3.443	3.049	2.817	2.661	2.549	2.464	2.397	2.297
23	4.279	3.422	3.028	2.796	2.640	2.528	2.442	2.375	2.275
24	4.260	3.403	3.009	2.776	2.621	2.508	2.423	2.355	2.255
25	4.242	3.385	2.991	2.759	2.603	2.490	2.405	2.337	2.236
26	4.225	3.369	2.975	2.743	2.587	2.474	2.388	2.321	2.220
27	4.210	3.354	2.960	2.728	2.572	2.459	2.373	2.305	2.204
28	4.196	3.340	2.947	2.714	2.558	2.445	2.359	2.291	2.190
29	4.183	3.328	2.934	2.701	2.545	2.432	2.346	2.278	2.177
30	4.171	3.316	2.922	2.690	2.534	2.421	2.334	2.266	2.165
35	4.121	3.267	2.874	2.641	2.485	2.372	2.285	2.217	2.114
40	4.085	3.232	2.839	2.606	2.449	2.336	2.249	2.180	2.077

(continued)

df2\df1	1	2	3	4	5	6	7	8	10
45	4.057	3.204	2.812	2.579	2.422	2.308	2.221	2.152	2.049
50	4.034	3.183	2.790	2.557	2.400	2.286	2.199	2.130	2.026
55	4.016	3.165	2.773	2.540	2.383	2.269	2.181	2.112	2.008
60	4.001	3.150	2.758	2.525	2.368	2.254	2.167	2.097	1.993
70	3.978	3.128	2.736	2.503	2.346	2.231	2.143	2.074	1.969
80	3.960	3.111	2.719	2.486	2.329	2.214	2.126	2.056	1.951
90	3.947	3.098	2.706	2.473	2.316	2.201	2.113	2.043	1.938
100	3.936	3.087	2.696	2.463	2.305	2.191	2.103	2.032	1.927
110	3.927	3.079	2.687	2.454	2.297	2.182	2.094	2.024	1.918
120	3.920	3.072	2.680	2.447	2.290	2.175	2.087	2.016	1.910
130	3.914	3.066	2.674	2.441	2.284	2.169	2.081	2.010	1.904
140	3.909	3.061	2.669	2.436	2.279	2.164	2.076	2.005	1.899
150	3.904	3.056	2.665	2.432	2.274	2.160	2.071	2.001	1.894
160	3.900	3.053	2.661	2.428	2.271	2.156	2.067	1.997	1.890
180	3.894	3.046	2.655	2.422	2.264	2.149	2.061	1.990	1.884
200	3.888	3.041	2.650	2.417	2.259	2.144	2.056	1.985	1.878
220	3.884	3.037	2.646	2.413	2.255	2.140	2.051	1.981	1.874
240	3.880	3.033	2.642	2.409	2.252	2.136	2.048	1.977	1.870
260	3.877	3.031	2.639	2.406	2.249	2.134	2.045	1.974	1.867
280	3.875	3.028	2.637	2.404	2.246	2.131	2.042	1.972	1.865
300	3.873	3.026	2.635	2.402	2.244	2.129	2.040	1.969	1.862
400	3.865	3.018	2.627	2.394	2.237	2.121	2.032	1.962	1.854
500	3.860	3.014	2.623	2.390	2.232	2.117	2.028	1.957	1.850
600	3.857	3.011	2.620	2.387	2.229	2.114	2.025	1.954	1.846
700	3.855	3.009	2.618	2.385	2.227	2.112	2.023	1.952	1.844
800	3.853	3.007	2.616	2.383	2.225	2.110	2.021	1.950	1.843
900	3.852	3.006	2.615	2.382	2.224	2.109	2.020	1.949	1.841
1000	3.851	3.005	2.614	2.381	2.223	2.108	2.019	1.948	1.840
∞	3.841	2.996	2.605	2.372	2.214	2.099	2.010	1.938	1.831

F-distribution (Upper tail probability = 0.05) Numerator df = 12 to 40

df2\df1	12	14	16	18	20	24	28	32	36	40
1	243.906	245.364	246.464	247.323	248.013	249.052	249.797	250.357	250.793	251.143
2	19.413	19.424	19.433	19.440	19.446	19.454	19.460	19.464	19.468	19.471
3	8.745	8.715	8.692	8.675	8.660	8.639	8.623	8.611	8.602	8.594
4	5.912	5.873	5.844	5.821	5.803	5.774	5.754	5.739	5.727	5.717
5	4.678	4.636	4.604	4.579	4.558	4.527	4.505	4.488	4.474	4.464
6	4.000	3.956	3.922	3.896	3.874	3.841	3.818	3.800	3.786	3.774
7	3.575	3.529	3.494	3.467	3.445	3.410	3.386	3.367	3.352	3.340
8	3.284	3.237	3.202	3.173	3.150	3.115	3.090	3.070	3.055	3.043
9	3.073	3.025	2.989	2.960	2.936	2.900	2.874	2.854	2.839	2.826
10	2.913	2.865	2.828	2.798	2.774	2.737	2.710	2.690	2.674	2.661
11	2.788	2.739	2.701	2.671	2.646	2.609	2.582	2.561	2.544	2.531
12	2.687	2.637	2.599	2.568	2.544	2.505	2.478	2.456	2.439	2.426
13	2.604	2.554	2.515	2.484	2.459	2.420	2.392	2.370	2.353	2.339
14	2.534	2.484	2.445	2.413	2.388	2.349	2.320	2.298	2.280	2.266
15	2.475	2.424	2.385	2.353	2.328	2.288	2.259	2.236	2.219	2.204
16	2.425	2.373	2.333	2.302	2.276	2.235	2.206	2.183	2.165	2.151
17	2.381	2.329	2.289	2.257	2.230	2.190	2.160	2.137	2.119	2.104
18	2.342	2.290	2.250	2.217	2.191	2.150	2.119	2.096	2.078	2.063
19	2.308	2.256	2.215	2.182	2.155	2.114	2.084	2.060	2.042	2.026
20	2.278	2.225	2.184	2.151	2.124	2.082	2.052	2.028	2.009	1.994
21	2.250	2.197	2.156	2.123	2.096	2.054	2.023	1.999	1.980	1.965
22	2.226	2.173	2.131	2.098	2.071	2.028	1.997	1.973	1.954	1.938
23	2.204	2.150	2.109	2.075	2.048	2.005	1.973	1.949	1.930	1.914
24	2.183	2.130	2.088	2.054	2.027	1.984	1.952	1.927	1.908	1.892
25	2.165	2.111	2.069	2.035	2.007	1.964	1.932	1.908	1.888	1.872
26	2.148	2.094	2.052	2.018	1.990	1.946	1.914	1.889	1.869	1.853
27	2.132	2.078	2.036	2.002	1.974	1.930	1.898	1.872	1.852	1.836
28	2.118	2.064	2.021	1.987	1.959	1.915	1.882	1.857	1.837	1.820
29	2.104	2.050	2.007	1.973	1.945	1.901	1.868	1.842	1.822	1.806
30	2.092	2.037	1.995	1.960	1.932	1.887	1.854	1.829	1.808	1.792
35	2.041	1.986	1.942	1.907	1.878	1.833	1.799	1.773	1.752	1.735
40	2.003	1.948	1.904	1.868	1.839	1.793	1.759	1.732	1.710	1.693
45	1.974	1.918	1.874	1.838	1.808	1.762	1.727	1.700	1.678	1.660

(continued)

df2\df1	12	14	16	18	20	24	28	32	36	40
50	1.952	1.895	1.850	1.814	1.784	1.737	1.702	1.674	1.652	1.634
55	1.933	1.876	1.831	1.795	1.764	1.717	1.681	1.653	1.631	1.612
60	1.917	1.860	1.815	1.778	1.748	1.700	1.664	1.636	1.613	1.594
70	1.893	1.836	1.790	1.753	1.722	1.674	1.637	1.608	1.585	1.566
80	1.875	1.817	1.772	1.734	1.703	1.654	1.617	1.588	1.564	1.545
90	1.861	1.803	1.757	1.720	1.688	1.639	1.601	1.572	1.548	1.528
100	1.850	1.792	1.746	1.708	1.676	1.627	1.589	1.559	1.535	1.515
110	1.841	1.783	1.736	1.698	1.667	1.617	1.579	1.549	1.524	1.504
120	1.834	1.775	1.728	1.690	1.659	1.608	1.570	1.540	1.516	1.495
130	1.827	1.769	1.722	1.684	1.652	1.601	1.563	1.533	1.508	1.488
140	1.822	1.763	1.716	1.678	1.646	1.595	1.557	1.526	1.502	1.481
150	1.817	1.758	1.711	1.673	1.641	1.590	1.552	1.521	1.496	1.475
160	1.813	1.754	1.707	1.669	1.637	1.586	1.547	1.516	1.491	1.470
180	1.806	1.747	1.700	1.661	1.629	1.578	1.539	1.508	1.483	1.462
200	1.801	1.742	1.694	1.656	1.623	1.572	1.533	1.502	1.476	1.455
220	1.796	1.737	1.690	1.651	1.618	1.567	1.528	1.496	1.471	1.450
240	1.793	1.733	1.686	1.647	1.614	1.563	1.523	1.492	1.466	1.445
260	1.790	1.730	1.683	1.644	1.611	1.559	1.520	1.488	1.463	1.441
280	1.787	1.727	1.680	1.641	1.608	1.556	1.517	1.485	1.459	1.438
300	1.785	1.725	1.677	1.638	1.606	1.554	1.514	1.482	1.456	1.435
400	1.776	1.717	1.669	1.630	1.597	1.545	1.505	1.473	1.447	1.425
500	1.772	1.712	1.664	1.625	1.592	1.539	1.499	1.467	1.441	1.419
600	1.768	1.708	1.660	1.621	1.588	1.536	1.495	1.463	1.437	1.414
700	1.766	1.706	1.658	1.619	1.586	1.533	1.492	1.460	1.434	1.412
800	1.764	1.704	1.656	1.617	1.584	1.531	1.490	1.458	1.432	1.409
900	1.763	1.703	1.655	1.615	1.582	1.529	1.489	1.457	1.430	1.408
1000	1.762	1.702	1.654	1.614	1.581	1.528	1.488	1.455	1.429	1.406
∞	1.752	1.692	1.644	1.604	1.571	1.517	1.476	1.444	1.417	1.394

Cumulative Binomial Distribution – 1

							p					
n	x	.01	.05	.10	.15	.20	.25	.30	.35	.40	.45	.50
2	0	0.9801	0.9025	0.8100	0.7225	0.6400	0.5625	0.4900	0.4225	0.3600	0.3025	0.2500
	1	0.9999	0.9975	0.9900	0.9775	0.9600	0.9375	0.9100	0.8775	0.8400	0.7975	0.7500
	2	1.0000	1.0000	1.0000	1.0000	1.0000	1.0000	1.0000	1.0000	1.0000	1.0000	1.0000
3	0	0.97030	0.85738	0.729	0.61413	0.512	0.42187	0.343	0.27463	0.216	0.16638	0.125
	1	0.99970	0.99275	0.972	0.93925	0.896	0.84375	0.784	0.71825	0.648	0.57475	0.500
	2	1.00000	0.99988	0.999	0.99663	0.992	0.98437	0.973	0.95713	0.936	0.90887	0.875
	3	1.00000	1.00000	1.000	1.00000	1.000	1.00000	1.000	1.00000	1.000	1.00000	1.000
4	0	0.96060	0.81451	0.6561	0.52201	0.4096	0.31641	0.2401	0.17851	0.1296	0.09151	0.0625
	1	0.99941	0.98598	0.9477	0.89048	0.8192	0.73828	0.6517	0.56298	0.4752	0.39098	0.3125
	2	1.00000	0.99952	0.9963	0.98802	0.9728	0.94922	0.9163	0.87352	0.8208	0.75852	0.6875
	3	1.00000	0.99999	0.9999	0.99949	0.9984	0.99609	0.9919	0.98499	0.9744	0.95899	0.9375
	4	1.00000	1.00000	1.0000	1.00000	1.0000	1.00000	1.0000	1.00000	1.0000	1.00000	1.0000
5	0	0.95099	0.77378	0.59049	0.44371	0.32768	0.23730	0.16807	0.11603	0.07776	0.05033	0.03125
	1	0.99902	0.97741	0.91854	0.83521	0.73728	0.63281	0.52822	0.42842	0.33696	0.25622	0.18750
	2	0.99999	0.99884	0.99144	0.97339	0.94208	0.89648	0.83692	0.76483	0.68256	0.59313	0.50000
	3	1.00000	0.99997	0.99954	0.99777	0.99328	0.98437	0.96922	0.94598	0.91296	0.86878	0.81250
	4	1.00000	1.00000	0.99999	0.99992	0.99968	0.99902	0.99757	0.99475	0.98976	0.98155	0.96875
	5	1.00000	1.00000	1.00000	1.00000	1.00000	1.00000	1.00000	1.00000	1.00000	1.00000	1.00000
6	0	0.94148	0.73509	0.53144	0.37715	0.26214	0.17798	0.11765	0.07542	0.04666	0.02768	0.01563
	1	0.99854	0.96723	0.88573	0.77648	0.65536	0.53394	0.42017	0.31908	0.23328	0.16357	0.10938
	2	0.99998	0.99777	0.98415	0.95266	0.90112	0.83057	0.74431	0.64709	0.54432	0.44152	0.34375
	3	1.00000	0.99991	0.99873	0.99411	0.98304	0.96240	0.92953	0.88258	0.82080	0.74474	0.65625
	4	1.00000	1.00000	0.99994	0.99960	0.99840	0.99536	0.98906	0.97768	0.95904	0.93080	0.89062
	5	1.00000	1.00000	1.00000	0.99999	0.99994	0.99976	0.99927	0.99816	0.99590	0.99170	0.98437
	6	1.00000	1.00000	1.00000	1.00000	1.00000	1.00000	1.00000	1.00000	1.00000	1.00000	1.00000
7	0	0.93207	0.69834	0.47830	0.32058	0.20972	0.13348	0.08235	0.04902	0.02799	0.01522	0.00781
	1	0.99797	0.95562	0.85031	0.71658	0.57672	0.44495	0.32942	0.23380	0.15863	0.10242	0.06250
	2	0.99997	0.99624	0.97431	0.92623	0.85197	0.75641	0.64707	0.53228	0.41990	0.31644	0.22656
	3	1.00000	0.99981	0.99727	0.98790	0.96666	0.92944	0.87396	0.80015	0.71021	0.60829	0.50000
	4	1.00000	0.99999	0.99982	0.99878	0.99533	0.98712	0.97120	0.94439	0.90374	0.84707	0.77344
	5	1.00000	1.00000	0.99999	0.99993	0.99963	0.99866	0.99621	0.99099	0.98116	0.96429	0.93750
	6	1.00000	1.00000	1.00000	1.00000	0.99999	0.99994	0.99978	0.99936	0.99836	0.99626	0.99219
	7	1.00000	1.00000	1.00000	1.00000	1.00000	1.00000	1.00000	1.00000	1.00000	1.00000	1.00000

(continued)

n	x	.01	.05	.10	.15	.20	.25	.30	.35	.40	.45	.50
8	0	0.92274	0.66342	0.43047	0.27249	0.16777	0.10011	0.05765	0.03186	0.01680	0.00837	0.00391
	1	0.99731	0.94276	0.81310	0.65718	0.50332	0.36708	0.25530	0.16913	0.10638	0.06318	0.03516
	2	0.99995	0.99421	0.96191	0.89479	0.79692	0.67854	0.55177	0.42781	0.31539	0.22013	0.14453
	3	1.00000	0.99963	0.99498	0.97865	0.94372	0.88618	0.80590	0.70640	0.59409	0.47696	0.36328
	4	1.00000	0.99998	0.99957	0.99715	0.98959	0.97270	0.94203	0.89391	0.82633	0.73962	0.63672
	5	1.00000	1.00000	0.99998	0.99976	0.99877	0.99577	0.98871	0.97468	0.95019	0.91154	0.85547
	6	1.00000	1.00000	1.00000	0.99999	0.99992	0.99962	0.99871	0.99643	0.99148	0.98188	0.96484
	7	1.00000	1.00000	1.00000	1.00000	1.00000	0.99998	0.99993	0.99977	0.99934	0.99832	0.99609
	8	1.00000	1.00000	1.00000	1.00000	1.00000	1.00000	1.00000	1.00000	1.00000	1.00000	1.00000

Cumulative Binomial Distribution – 2

n	x	.01	.05	.10	.15	.20	.25	.30	.35	.40	.45	.50
9	0	0.91352	0.63025	0.38742	0.23162	0.13422	0.07508	0.04035	0.02071	0.01008	0.00461	0.00195
	1	0.99656	0.92879	0.77484	0.59948	0.43621	0.30034	0.19600	0.12109	0.07054	0.03852	0.01953
	2	0.99992	0.99164	0.94703	0.85915	0.73820	0.60068	0.46283	0.33727	0.23179	0.14950	0.08984
	3	1.00000	0.99936	0.99167	0.96607	0.91436	0.83427	0.72966	0.60889	0.48261	0.36138	0.25391
	4	1.00000	0.99997	0.99911	0.99437	0.98042	0.95107	0.90119	0.82828	0.73343	0.62142	0.50000
	5	1.00000	1.00000	0.99994	0.99937	0.99693	0.99001	0.97471	0.94641	0.90065	0.83418	0.74609
	6	1.00000	1.00000	1.00000	0.99995	0.99969	0.99866	0.99571	0.98882	0.97497	0.95023	0.91016
	7	1.00000	1.00000	1.00000	1.00000	0.99998	0.99989	0.99957	0.99860	0.99620	0.99092	0.98047
	8	1.00000	1.00000	1.00000	1.00000	1.00000	1.00000	0.99998	0.99992	0.99974	0.99924	0.99805
	9	1.00000	1.00000	1.00000	1.00000	1.00000	1.00000	1.00000	1.00000	1.00000	1.00000	1.00000

(continued)

n	x	.01	.05	.10	.15	.20	.25	.30	.35	.40	.45	.50
10	0	0.90438	0.59874	0.34868	0.19687	0.10737	0.05631	0.02825	0.01346	0.00605	0.00253	0.00098
	1	0.99573	0.91386	0.73610	0.54430	0.37581	0.24403	0.14931	0.08595	0.04636	0.02326	0.01074
	2	0.99989	0.98850	0.92981	0.82020	0.67780	0.52559	0.38278	0.26161	0.16729	0.09956	0.05469
	3	1.00000	0.99897	0.98720	0.95003	0.87913	0.77588	0.64961	0.51383	0.38228	0.26604	0.17188
	4	1.00000	0.99994	0.99837	0.99013	0.96721	0.92187	0.84973	0.75150	0.63310	0.50440	0.37695
	5	1.00000	1.00000	0.99985	0.99862	0.99363	0.98027	0.95265	0.90507	0.83376	0.73844	0.62305
	6	1.00000	1.00000	0.99999	0.99987	0.99914	0.99649	0.98941	0.97398	0.94524	0.89801	0.82812
	7	1.00000	1.00000	1.00000	0.99999	0.99992	0.99958	0.99841	0.99518	0.98771	0.97261	0.94531
	8	1.00000	1.00000	1.00000	1.00000	1.00000	0.99997	0.99986	0.99946	0.99832	0.99550	0.98926
	9	1.00000	1.00000	1.00000	1.00000	1.00000	1.00000	0.99999	0.99997	0.99990	0.99966	0.99902
	10	1.00000	1.00000	1.00000	1.00000	1.00000	1.00000	1.00000	1.00000	1.00000	1.00000	1.00000
11	0	0.89534	0.56880	0.31381	0.16734	0.08590	0.04224	0.01977	0.00875	0.00363	0.00139	0.00049
	1	0.99482	0.89811	0.69736	0.49219	0.32212	0.19710	0.11299	0.06058	0.03023	0.01393	0.00586
	2	0.99984	0.98476	0.91044	0.77881	0.61740	0.45520	0.31274	0.20013	0.11892	0.06522	0.03271
	3	1.00000	0.99845	0.98147	0.93056	0.83886	0.71330	0.56956	0.42555	0.29628	0.19112	0.11328
	4	1.00000	0.99989	0.99725	0.98411	0.94959	0.88537	0.78970	0.66831	0.53277	0.39714	0.27441
	5	1.00000	0.99999	0.99970	0.99734	0.98835	0.96567	0.92178	0.85132	0.75350	0.63312	0.50000
	6	1.00000	1.00000	0.99998	0.99968	0.99803	0.99244	0.97838	0.94986	0.90065	0.82620	0.72559
	7	1.00000	1.00000	1.00000	0.99997	0.99976	0.99881	0.99571	0.98776	0.97072	0.93904	0.88672
	8	1.00000	1.00000	1.00000	1.00000	0.99998	0.99987	0.99942	0.99796	0.99408	0.98520	0.96729
	9	1.00000	1.00000	1.00000	1.00000	1.00000	0.99999	0.99995	0.99979	0.99927	0.99779	0.99414
	10	1.00000	1.00000	1.00000	1.00000	1.00000	1.00000	1.00000	0.99999	0.99996	0.99985	0.99951
	11	1.00000	1.00000	1.00000	1.00000	1.00000	1.00000	1.00000	1.00000	1.00000	1.00000	1.00000
12	0	0.88638	0.54036	0.28243	0.14224	0.06872	0.03168	0.01384	0.00569	0.00218	0.00077	0.00024
	1	0.99383	0.88164	0.65900	0.44346	0.27488	0.15838	0.08503	0.04244	0.01959	0.00829	0.00317
	2	0.99979	0.98043	0.88913	0.73582	0.55835	0.39068	0.25282	0.15129	0.08344	0.04214	0.01929
	3	1.00000	0.99776	0.97436	0.90779	0.79457	0.64878	0.49252	0.34665	0.22534	0.13447	0.07300
	4	1.00000	0.99982	0.99567	0.97608	0.92744	0.84236	0.72366	0.58335	0.43818	0.30443	0.19385
	5	1.00000	0.99999	0.99946	0.99536	0.98059	0.94560	0.88215	0.78726	0.66521	0.52693	0.38721
	6	1.00000	1.00000	0.99995	0.99933	0.99610	0.98575	0.96140	0.91537	0.84179	0.73931	0.61279
	7	1.00000	1.00000	1.00000	0.99993	0.99942	0.99722	0.99051	0.97449	0.94269	0.88826	0.80615
	8	1.00000	1.00000	1.00000	0.99999	0.99994	0.99961	0.99831	0.99439	0.98473	0.96443	0.92700

Cumulative Binomial Distribution – 3

							p					
n	x	.01	.05	.10	.15	.20	.25	.30	.35	.40	.45	.50
12	9	1.00000	1.00000	1.00000	1.00000	1.00000	0.99996	0.99979	0.99915	0.99719	0.99212	0.98071
	10	1.00000	1.00000	1.00000	1.00000	1.00000	1.00000	0.99998	0.99992	0.99968	0.99892	0.99683
	11	1.00000	1.00000	1.00000	1.00000	1.00000	1.00000	1.00000	1.00000	0.99998	0.99993	0.99976
	12	1.00000	1.00000	1.00000	1.00000	1.00000	1.00000	1.00000	1.00000	1.00000	1.00000	1.00000
13	0	0.87752	0.51334	0.25419	0.12091	0.05498	0.02376	0.00969	0.00370	0.00131	0.00042	0.00012
	1	0.99275	0.86458	0.62134	0.39828	0.23365	0.12671	0.06367	0.02958	0.01263	0.00490	0.00171
	2	0.99973	0.97549	0.86612	0.69196	0.50165	0.33260	0.20248	0.11319	0.05790	0.02691	0.01123
	3	0.99999	0.99690	0.96584	0.88200	0.74732	0.58425	0.42061	0.27827	0.16858	0.09292	0.04614
	4	1.00000	0.99971	0.99354	0.96584	0.90087	0.79396	0.65431	0.50050	0.35304	0.22795	0.13342
	5	1.00000	0.99998	0.99908	0.99247	0.96996	0.91979	0.83460	0.71589	0.57440	0.42681	0.29053
	6	1.00000	1.00000	0.99990	0.99873	0.99300	0.97571	0.93762	0.87053	0.77116	0.64374	0.50000
	7	1.00000	1.00000	0.99999	0.99984	0.99875	0.99435	0.98178	0.95380	0.90233	0.82123	0.70947
	8	1.00000	1.00000	1.00000	0.99998	0.99983	0.99901	0.99597	0.98743	0.96792	0.93015	0.86658
	9	1.00000	1.00000	1.00000	1.00000	0.99998	0.99987	0.99935	0.99749	0.99221	0.97966	0.95386
	10	1.00000	1.00000	1.00000	1.00000	1.00000	0.99999	0.99993	0.99965	0.99868	0.99586	0.98877
	11	1.00000	1.00000	1.00000	1.00000	1.00000	1.00000	0.99999	0.99997	0.99986	0.99948	0.99829
	12	1.00000	1.00000	1.00000	1.00000	1.00000	1.00000	1.00000	1.00000	0.99999	0.99997	0.99988
	13	1.00000	1.00000	1.00000	1.00000	1.00000	1.00000	1.00000	1.00000	1.00000	1.00000	1.00000
14	0	0.86875	0.48767	0.22877	0.10277	0.04398	0.01782	0.00678	0.00240	0.00078	0.00023	0.00006
	1	0.99160	0.84701	0.58463	0.35667	0.19791	0.10097	0.04748	0.02052	0.00810	0.00289	0.00092
	2	0.99966	0.96995	0.84164	0.64791	0.44805	0.28113	0.16084	0.08393	0.03979	0.01701	0.00647
	3	0.99999	0.99583	0.95587	0.85349	0.69819	0.52134	0.35517	0.22050	0.12431	0.06322	0.02869
	4	1.00000	0.99957	0.99077	0.95326	0.87016	0.74153	0.58420	0.42272	0.27926	0.16719	0.08978
	5	1.00000	0.99997	0.99853	0.98847	0.95615	0.88833	0.78052	0.64051	0.48585	0.33732	0.21198
	6	1.00000	1.00000	0.99982	0.99779	0.98839	0.96173	0.90672	0.81641	0.69245	0.54612	0.39526
	7	1.00000	1.00000	0.99998	0.99967	0.99760	0.98969	0.96853	0.92466	0.84986	0.74136	0.60474
	8	1.00000	1.00000	1.00000	0.99996	0.99962	0.99785	0.99171	0.97566	0.94168	0.88114	0.78802
	9	1.00000	1.00000	1.00000	1.00000	0.99995	0.99966	0.99833	0.99396	0.98249	0.95738	0.91022
	10	1.00000	1.00000	1.00000	1.00000	1.00000	0.99996	0.99975	0.99889	0.99609	0.98857	0.97131
	11	1.00000	1.00000	1.00000	1.00000	1.00000	1.00000	0.99997	0.99986	0.99939	0.99785	0.99353
	12	1.00000	1.00000	1.00000	1.00000	1.00000	1.00000	1.00000	0.99999	0.99994	0.99975	0.99908
	13	1.00000	1.00000	1.00000	1.00000	1.00000	1.00000	1.00000	1.00000	1.00000	0.99999	0.99994
	14	1.00000	1.00000	1.00000	1.00000	1.00000	1.00000	1.00000	1.00000	1.00000	1.00000	1.00000

Cumulative Binomial Distribution – 4

n	x	.01	.05	.10	.15	.20	.25	.30	.35	.40	.45	.50
							p					
15	0	0.86006	0.46329	0.20589	0.08735	0.03518	0.01336	0.00475	0.00156	0.00047	0.00013	0.00003
	1	0.99037	0.82905	0.54904	0.31859	0.16713	0.08018	0.03527	0.01418	0.00517	0.00169	0.00049
	2	0.99958	0.96380	0.81594	0.60423	0.39802	0.23609	0.12683	0.06173	0.02711	0.01065	0.00369
	3	0.99999	0.99453	0.94444	0.82266	0.64816	0.46129	0.29687	0.17270	0.09050	0.04242	0.01758
	4	1.00000	0.99939	0.98728	0.93829	0.83577	0.68649	0.51549	0.35194	0.21728	0.12040	0.05923
	5	1.00000	0.99995	0.99775	0.98319	0.93895	0.85163	0.72162	0.56428	0.40322	0.26076	0.15088
	6	1.00000	1.00000	0.99969	0.99639	0.98194	0.94338	0.86886	0.75484	0.60981	0.45216	0.30362
	7	1.00000	1.00000	0.99997	0.99939	0.99576	0.98270	0.94999	0.88677	0.78690	0.65350	0.50000
	8	1.00000	1.00000	1.00000	0.99992	0.99922	0.99581	0.98476	0.95781	0.90495	0.81824	0.69638
	9	1.00000	1.00000	1.00000	0.99999	0.99989	0.99921	0.99635	0.98756	0.96617	0.92307	0.84912
	10	1.00000	1.00000	1.00000	1.00000	0.99999	0.99988	0.99933	0.99717	0.99065	0.97453	0.94077
	11	1.00000	1.00000	1.00000	1.00000	1.00000	0.99999	0.99991	0.99952	0.99807	0.99367	0.98242
	12	1.00000	1.00000	1.00000	1.00000	1.00000	1.00000	0.99999	0.99994	0.99972	0.99889	0.99631
	13	1.00000	1.00000	1.00000	1.00000	1.00000	1.00000	1.00000	1.00000	0.99997	0.99988	0.99951
	14	1.00000	1.00000	1.00000	1.00000	1.00000	1.00000	1.00000	1.00000	1.00000	0.99999	0.99997
	15	1.00000	1.00000	1.00000	1.00000	1.00000	1.00000	1.00000	1.00000	1.00000	1.00000	1.00000

(continued)

n	x	.01	.05	.10	.15	.20	.25	p .30	.35	.40	.45	.50
20	0	0.81791	0.35849	0.12158	0.03876	0.01153	0.00317	0.00080	0.00018	0.00004	0.00001	0.00000
	1	0.98314	0.73584	0.39175	0.17556	0.06918	0.02431	0.00764	0.00213	0.00052	0.00011	0.00002
	2	0.99900	0.92452	0.67693	0.40490	0.20608	0.09126	0.03548	0.01212	0.00361	0.00093	0.00020
	3	0.99996	0.98410	0.86705	0.64773	0.41145	0.22516	0.10709	0.04438	0.01596	0.00493	0.00129
	4	1.00000	0.99743	0.95683	0.82985	0.62965	0.41484	0.23751	0.11820	0.05095	0.01886	0.00591
	5	1.00000	0.99967	0.98875	0.93269	0.80421	0.61717	0.41637	0.24540	0.12560	0.05533	0.02069
	6	1.00000	0.99997	0.99761	0.97806	0.91331	0.78578	0.60801	0.41663	0.25001	0.12993	0.05766
	7	1.00000	1.00000	0.99958	0.99408	0.96786	0.89819	0.77227	0.60103	0.41589	0.25201	0.13159
	8	1.00000	1.00000	0.99994	0.99867	0.99002	0.95907	0.88667	0.76238	0.59560	0.41431	0.25172
	9	1.00000	1.00000	0.99999	0.99975	0.99741	0.98614	0.95204	0.87822	0.75534	0.59136	0.41190
	10	1.00000	1.00000	1.00000	0.99996	0.99944	0.99606	0.98286	0.94683	0.87248	0.75071	0.58810
	11	1.00000	1.00000	1.00000	0.99999	0.99990	0.99906	0.99486	0.98042	0.94347	0.86923	0.74828
	12	1.00000	1.00000	1.00000	1.00000	0.99998	0.99982	0.99872	0.99398	0.97897	0.94197	0.86841
	13	1.00000	1.00000	1.00000	1.00000	1.00000	0.99997	0.99974	0.99848	0.99353	0.97859	0.94234
	14	1.00000	1.00000	1.00000	1.00000	1.00000	1.00000	0.99996	0.99969	0.99839	0.99357	0.97931
	15	1.00000	1.00000	1.00000	1.00000	1.00000	1.00000	0.99999	0.99995	0.99968	0.99847	0.99409
	16	1.00000	1.00000	1.00000	1.00000	1.00000	1.00000	1.00000	0.99999	0.99995	0.99972	0.99871
	17	1.00000	1.00000	1.00000	1.00000	1.00000	1.00000	1.00000	1.00000	0.99999	0.99996	0.99980
	18	1.00000	1.00000	1.00000	1.00000	1.00000	1.00000	1.00000	1.00000	1.00000	1.00000	0.99998
	19	1.00000	1.00000	1.00000	1.00000	1.00000	1.00000	1.00000	1.00000	1.00000	1.00000	1.00000
	20	1.00000	1.00000	1.00000	1.00000	1.00000	1.00000	1.00000	1.00000	1.00000	1.00000	1.00000

Cumulative Binomial Distribution – 5

							p					
n	x	.01	.05	.10	.15	.20	.25	.30	.35	.40	.45	.50
25	0	0.77782	0.27739	0.07179	0.01720	0.00378	0.00075	0.00013	0.00002	0.00000	0.00000	0.00000
	1	0.97424	0.64238	0.27121	0.09307	0.02739	0.00702	0.00157	0.00030	0.00005	0.00001	0.00000
	2	0.99805	0.87289	0.53709	0.25374	0.09823	0.03211	0.00896	0.00213	0.00043	0.00007	0.00001
	3	0.99989	0.96591	0.76359	0.47112	0.23399	0.09621	0.03324	0.00968	0.00237	0.00048	0.00008
	4	1.00000	0.99284	0.90201	0.68211	0.42067	0.21374	0.09047	0.03205	0.00947	0.00231	0.00046
	5	1.00000	0.99879	0.96660	0.83848	0.61669	0.37828	0.19349	0.08262	0.02936	0.00860	0.00204
	6	1.00000	0.99983	0.99052	0.93047	0.78004	0.56110	0.34065	0.17340	0.07357	0.02575	0.00732
	7	1.00000	0.99998	0.99774	0.97453	0.89088	0.72651	0.51185	0.30608	0.15355	0.06385	0.02164
	8	1.00000	1.00000	0.99954	0.99203	0.95323	0.85056	0.67693	0.46682	0.27353	0.13398	0.05388
	9	1.00000	1.00000	0.99992	0.99786	0.98267	0.92867	0.81056	0.63031	0.42462	0.24237	0.11476
	10	1.00000	1.00000	0.99999	0.99951	0.99445	0.97033	0.90220	0.77116	0.58577	0.38426	0.21218
	11	1.00000	1.00000	1.00000	0.99990	0.99846	0.98927	0.95575	0.87458	0.73228	0.54257	0.34502
	12	1.00000	1.00000	1.00000	0.99998	0.99963	0.99663	0.98253	0.93956	0.84623	0.69368	0.50000
	13	1.00000	1.00000	1.00000	1.00000	0.99992	0.99908	0.99401	0.97454	0.92220	0.81731	0.65498
	14	1.00000	1.00000	1.00000	1.00000	0.99999	0.99979	0.99822	0.99069	0.96561	0.90402	0.78782
	15	1.00000	1.00000	1.00000	1.00000	1.00000	0.99996	0.99955	0.99706	0.98683	0.95604	0.88524
	16	1.00000	1.00000	1.00000	1.00000	1.00000	0.99999	0.99990	0.99921	0.99567	0.98264	0.94612
	17	1.00000	1.00000	1.00000	1.00000	1.00000	1.00000	0.99998	0.99982	0.99879	0.99417	0.97836
	18	1.00000	1.00000	1.00000	1.00000	1.00000	1.00000	1.00000	0.99997	0.99972	0.99836	0.99268
	19	1.00000	1.00000	1.00000	1.00000	1.00000	1.00000	1.00000	0.99999	0.99995	0.99962	0.99796
	20	1.00000	1.00000	1.00000	1.00000	1.00000	1.00000	1.00000	1.00000	0.99999	0.99993	0.99954
	21	1.00000	1.00000	1.00000	1.00000	1.00000	1.00000	1.00000	1.00000	1.00000	0.99999	0.99992
	22	1.00000	1.00000	1.00000	1.00000	1.00000	1.00000	1.00000	1.00000	1.00000	1.00000	0.99999
	23	1.00000	1.00000	1.00000	1.00000	1.00000	1.00000	1.00000	1.00000	1.00000	1.00000	1.00000
	24	1.00000	1.00000	1.00000	1.00000	1.00000	1.00000	1.00000	1.00000	1.00000	1.00000	1.00000
	25	1.00000	1.00000	1.00000	1.00000	1.00000	1.00000	1.00000	1.00000	1.00000	1.00000	1.00000

(continued)

							p					
n	x	.01	.05	.10	.15	.20	.25	.30	.35	.40	.45	.50
50	0	0.60501	0.07694	0.00515	0.00030	0.00001	0.00000	0.00000	0.00000	0.00000	0.00000	0.00000
	1	0.91056	0.27943	0.03379	0.00291	0.00019	0.00001	0.00000	0.00000	0.00000	0.00000	0.00000
	2	0.98618	0.54053	0.11173	0.01419	0.00129	0.00009	0.00000	0.00000	0.00000	0.00000	0.00000
	3	0.99840	0.76041	0.25029	0.04605	0.00566	0.00050	0.00003	0.00000	0.00000	0.00000	0.00000
	4	0.99985	0.89638	0.43120	0.11211	0.01850	0.00211	0.00017	0.00001	0.00000	0.00000	0.00000
	5	0.99999	0.96222	0.61612	0.21935	0.04803	0.00705	0.00072	0.00005	0.00000	0.00000	0.00000
	6	1.00000	0.98821	0.77023	0.36130	0.10340	0.01939	0.00249	0.00022	0.00001	0.00000	0.00000
	7	1.00000	0.99681	0.87785	0.51875	0.19041	0.04526	0.00726	0.00080	0.00006	0.00000	0.00000
	8	1.00000	0.99924	0.94213	0.66810	0.30733	0.09160	0.01825	0.00248	0.00023	0.00001	0.00000
	9	1.00000	0.99984	0.97546	0.79109	0.44374	0.16368	0.04023	0.00670	0.00076	0.00006	0.00000
	10	1.00000	0.99997	0.99065	0.88008	0.58356	0.26220	0.07885	0.01601	0.00220	0.00020	0.00001
	11	1.00000	1.00000	0.99678	0.93719	0.71067	0.38162	0.13904	0.03423	0.00569	0.00063	0.00005
	12	1.00000	1.00000	0.99900	0.96994	0.81394	0.51099	0.22287	0.06613	0.01325	0.00177	0.00015
	13	1.00000	1.00000	0.99971	0.98683	0.88941	0.63704	0.32788	0.11633	0.02799	0.00449	0.00047
	14	1.00000	1.00000	0.99993	0.99471	0.93928	0.74808	0.44683	0.18778	0.05395	0.01038	0.00130
	15	1.00000	1.00000	0.99998	0.99805	0.96920	0.83692	0.56918	0.28010	0.09550	0.02195	0.00330
	16	1.00000	1.00000	1.00000	0.99934	0.98556	0.90169	0.68388	0.38886	0.15609	0.04265	0.00767
	17	1.00000	1.00000	1.00000	0.99979	0.99374	0.94488	0.78219	0.50597	0.23688	0.07653	0.01642
	18	1.00000	1.00000	1.00000	0.99994	0.99749	0.97127	0.85944	0.62159	0.33561	0.12734	0.03245
	19	1.00000	1.00000	1.00000	0.99998	0.99907	0.98608	0.91520	0.72644	0.44648	0.19737	0.05946
	20	1.00000	1.00000	1.00000	1.00000	0.99968	0.99374	0.95224	0.81394	0.56103	0.28617	0.10132

Cumulative Binomial Distribution – 6

							p					
n	x	.01	.05	.10	.15	.20	.25	.30	.35	.40	.45	.50
50	21	1.00000	1.00000	1.00000	1.00000	0.99990	0.99738	0.97491	0.88126	0.67014	0.38996	0.16112
ctd	22	1.00000	1.00000	1.00000	1.00000	0.99997	0.99898	0.98772	0.92904	0.76602	0.50191	0.23994
	23	1.00000	1.00000	1.00000	1.00000	0.99999	0.99963	0.99441	0.96036	0.84383	0.61341	0.33591
	24	1.00000	1.00000	1.00000	1.00000	1.00000	0.99988	0.99763	0.97933	0.90219	0.71604	0.44386
	25	1.00000	1.00000	1.00000	1.00000	1.00000	0.99996	0.99907	0.98996	0.94266	0.80337	0.55614
	26	1.00000	1.00000	1.00000	1.00000	1.00000	0.99999	0.99966	0.99546	0.96859	0.87207	0.66409
	27	1.00000	1.00000	1.00000	1.00000	1.00000	1.00000	0.99988	0.99809	0.98397	0.92204	0.76006
	28	1.00000	1.00000	1.00000	1.00000	1.00000	1.00000	0.99996	0.99925	0.99238	0.95562	0.83888
	29	1.00000	1.00000	1.00000	1.00000	1.00000	1.00000	0.99999	0.99973	0.99664	0.97646	0.89868
	30	1.00000	1.00000	1.00000	1.00000	1.00000	1.00000	1.00000	0.99991	0.99863	0.98840	0.94054
	31	1.00000	1.00000	1.00000	1.00000	1.00000	1.00000	1.00000	0.99997	0.99948	0.99470	0.96755
	32	1.00000	1.00000	1.00000	1.00000	1.00000	1.00000	1.00000	0.99999	0.99982	0.99776	0.98358
	33	1.00000	1.00000	1.00000	1.00000	1.00000	1.00000	1.00000	1.00000	0.99994	0.99913	0.99233
	34	1.00000	1.00000	1.00000	1.00000	1.00000	1.00000	1.00000	1.00000	0.99998	0.99969	0.99670
	35	1.00000	1.00000	1.00000	1.00000	1.00000	1.00000	1.00000	1.00000	1.00000	0.99990	0.99870
	36	1.00000	1.00000	1.00000	1.00000	1.00000	1.00000	1.00000	1.00000	1.00000	0.99997	0.99953
	37	1.00000	1.00000	1.00000	1.00000	1.00000	1.00000	1.00000	1.00000	1.00000	0.99999	0.99985
	38	1.00000	1.00000	1.00000	1.00000	1.00000	1.00000	1.00000	1.00000	1.00000	1.00000	0.99995
	39	1.00000	1.00000	1.00000	1.00000	1.00000	1.00000	1.00000	1.00000	1.00000	1.00000	0.99999
	40	1.00000	1.00000	1.00000	1.00000	1.00000	1.00000	1.00000	1.00000	1.00000	1.00000	1.00000
	41	1.00000	1.00000	1.00000	1.00000	1.00000	1.00000	1.00000	1.00000	1.00000	1.00000	1.00000
	42	1.00000	1.00000	1.00000	1.00000	1.00000	1.00000	1.00000	1.00000	1.00000	1.00000	1.00000
	43	1.00000	1.00000	1.00000	1.00000	1.00000	1.00000	1.00000	1.00000	1.00000	1.00000	1.00000
	44	1.00000	1.00000	1.00000	1.00000	1.00000	1.00000	1.00000	1.00000	1.00000	1.00000	1.00000
	45	1.00000	1.00000	1.00000	1.00000	1.00000	1.00000	1.00000	1.00000	1.00000	1.00000	1.00000
	46	1.00000	1.00000	1.00000	1.00000	1.00000	1.00000	1.00000	1.00000	1.00000	1.00000	1.00000
	47	1.00000	1.00000	1.00000	1.00000	1.00000	1.00000	1.00000	1.00000	1.00000	1.00000	1.00000
	48	1.00000	1.00000	1.00000	1.00000	1.00000	1.00000	1.00000	1.00000	1.00000	1.00000	1.00000
	49	1.00000	1.00000	1.00000	1.00000	1.00000	1.00000	1.00000	1.00000	1.00000	1.00000	1.00000
	50	1.00000	1.00000	1.00000	1.00000	1.00000	1.00000	1.00000	1.00000	1.00000	1.00000	1.00000

(continued)

n	x	.01	.05	.10	.15	.20	.25	.30	.35	.40	.45	.50
100	0	0.36603	0.00592	0.00003	0.00000	0.00000	0.00000	0.00000	0.00000	0.00000	0.00000	0.00000
	1	0.73576	0.03708	0.00032	0.00000	0.00000	0.00000	0.00000	0.00000	0.00000	0.00000	0.00000
	2	0.92063	0.11826	0.00194	0.00002	0.00000	0.00000	0.00000	0.00000	0.00000	0.00000	0.00000
	3	0.98163	0.25784	0.00784	0.00009	0.00000	0.00000	0.00000	0.00000	0.00000	0.00000	0.00000
	4	0.99657	0.43598	0.02371	0.00043	0.00000	0.00000	0.00000	0.00000	0.00000	0.00000	0.00000
	5	0.99947	0.61600	0.05758	0.00155	0.00002	0.00000	0.00000	0.00000	0.00000	0.00000	0.00000
	6	0.99993	0.76601	0.11716	0.00470	0.00008	0.00000	0.00000	0.00000	0.00000	0.00000	0.00000
	7	0.99999	0.87204	0.20605	0.01217	0.00028	0.00000	0.00000	0.00000	0.00000	0.00000	0.00000
	8	1.00000	0.93691	0.32087	0.02748	0.00086	0.00001	0.00000	0.00000	0.00000	0.00000	0.00000
	9	1.00000	0.97181	0.45129	0.05509	0.00233	0.00004	0.00000	0.00000	0.00000	0.00000	0.00000
	10	1.00000	0.98853	0.58316	0.09945	0.00570	0.00014	0.00000	0.00000	0.00000	0.00000	0.00000
	11	1.00000	0.99573	0.70303	0.16349	0.01257	0.00039	0.00001	0.00000	0.00000	0.00000	0.00000
	12	1.00000	0.99854	0.80182	0.24730	0.02533	0.00103	0.00002	0.00000	0.00000	0.00000	0.00000
	13	1.00000	0.99954	0.87612	0.34742	0.04691	0.00246	0.00006	0.00000	0.00000	0.00000	0.00000
	14	1.00000	0.99986	0.92743	0.45722	0.08044	0.00542	0.00016	0.00000	0.00000	0.00000	0.00000
	15	1.00000	0.99996	0.96011	0.56832	0.12851	0.01108	0.00040	0.00001	0.00000	0.00000	0.00000

Cumulative Binomial Distribution – 7

n	x	.01	.05	.10	.15	.20	.25	.30	.35	.40	.45	.50
100	16	1.00000	0.99999	0.97940	0.67246	0.19234	0.02111	0.00097	0.00002	0.00000	0.00000	0.00000
ctd	17	1.00000	1.00000	0.98999	0.76328	0.27119	0.03763	0.00216	0.00005	0.00000	0.00000	0.00000
	18	1.00000	1.00000	0.99542	0.83717	0.36209	0.06301	0.00452	0.00014	0.00000	0.00000	0.00000
	19	1.00000	1.00000	0.99802	0.89346	0.46016	0.09953	0.00889	0.00034	0.00001	0.00000	0.00000
	20	1.00000	1.00000	0.99919	0.93368	0.55946	0.14883	0.01646	0.00078	0.00002	0.00000	0.00000
	21	1.00000	1.00000	0.99969	0.96072	0.65403	0.21144	0.02883	0.00169	0.00004	0.00000	0.00000
	22	1.00000	1.00000	0.99989	0.97786	0.73893	0.28637	0.04787	0.00343	0.00011	0.00000	0.00000
	23	1.00000	1.00000	0.99996	0.98811	0.81091	0.37108	0.07553	0.00662	0.00025	0.00000	0.00000
	24	1.00000	1.00000	0.99999	0.99392	0.86865	0.46167	0.11357	0.01213	0.00056	0.00001	0.00000
	25	1.00000	1.00000	1.00000	0.99703	0.91252	0.55347	0.16313	0.02114	0.00119	0.00003	0.00000
	26	1.00000	1.00000	1.00000	0.99862	0.94417	0.64174	0.22440	0.03514	0.00240	0.00007	0.00000
	27	1.00000	1.00000	1.00000	0.99939	0.96585	0.72238	0.29637	0.05581	0.00460	0.00016	0.00000
	28	1.00000	1.00000	1.00000	0.99974	0.97998	0.79246	0.37678	0.08482	0.00843	0.00036	0.00001
	29	1.00000	1.00000	1.00000	0.99989	0.98875	0.85046	0.46234	0.12360	0.01478	0.00076	0.00002
	30	1.00000	1.00000	1.00000	0.99996	0.99394	0.89621	0.54912	0.17302	0.02478	0.00154	0.00004
	31	1.00000	1.00000	1.00000	0.99998	0.99687	0.93065	0.63311	0.23311	0.03985	0.00297	0.00009
	32	1.00000	1.00000	1.00000	0.99999	0.99845	0.95540	0.71072	0.30288	0.06150	0.00550	0.00020
	33	1.00000	1.00000	1.00000	1.00000	0.99926	0.97241	0.77926	0.38029	0.09125	0.00976	0.00044
	34	1.00000	1.00000	1.00000	1.00000	0.99966	0.98357	0.83714	0.46243	0.13034	0.01663	0.00089
	35	1.00000	1.00000	1.00000	1.00000	0.99985	0.99059	0.88392	0.54584	0.17947	0.02724	0.00176
	36	1.00000	1.00000	1.00000	1.00000	0.99994	0.99482	0.92012	0.62692	0.23861	0.04290	0.00332
	37	1.00000	1.00000	1.00000	1.00000	0.99998	0.99725	0.94695	0.70245	0.30681	0.06507	0.00602
	38	1.00000	1.00000	1.00000	1.00000	0.99999	0.99860	0.96602	0.76987	0.38219	0.09514	0.01049
	39	1.00000	1.00000	1.00000	1.00000	1.00000	0.99931	0.97901	0.82758	0.46208	0.13425	0.01760
	40	1.00000	1.00000	1.00000	1.00000	1.00000	0.99968	0.98750	0.87498	0.54329	0.18306	0.02844
	41	1.00000	1.00000	1.00000	1.00000	1.00000	0.99985	0.99283	0.91232	0.62253	0.24149	0.04431
	42	1.00000	1.00000	1.00000	1.00000	1.00000	0.99994	0.99603	0.94057	0.69674	0.30865	0.06661
	43	1.00000	1.00000	1.00000	1.00000	1.00000	0.99997	0.99789	0.96109	0.76347	0.38277	0.09667
	44	1.00000	1.00000	1.00000	1.00000	1.00000	0.99999	0.99891	0.97540	0.82110	0.46133	0.13563
	45	1.00000	1.00000	1.00000	1.00000	1.00000	1.00000	0.99946	0.98499	0.86891	0.54132	0.18410
	46	1.00000	1.00000	1.00000	1.00000	1.00000	1.00000	0.99974	0.99116	0.90702	0.61956	0.24206
	47	1.00000	1.00000	1.00000	1.00000	1.00000	1.00000	0.99988	0.99498	0.93621	0.69312	0.30865
	48	1.00000	1.00000	1.00000	1.00000	1.00000	1.00000	0.99995	0.99725	0.95770	0.75957	0.38218
	49	1.00000	1.00000	1.00000	1.00000	1.00000	1.00000	0.99998	0.99855	0.97290	0.81727	0.46020
	50	1.00000	1.00000	1.00000	1.00000	1.00000	1.00000	0.99999	0.99926	0.98324	0.86542	0.53979

(continued)

n	x	.01	.05	.10	.15	.20	.25	.30	.35	.40	.45	.50
	51	1.00000	1.00000	1.00000	1.00000	1.00000	1.00000	1.00000	0.99964	0.98999	0.90405	0.61782
	52	1.00000	1.00000	1.00000	1.00000	1.00000	1.00000	1.00000	0.99983	0.99424	0.93383	0.69135
	53	1.00000	1.00000	1.00000	1.00000	1.00000	1.00000	1.00000	0.99992	0.99680	0.95589	0.75794
	54	1.00000	1.00000	1.00000	1.00000	1.00000	1.00000	1.00000	0.99997	0.99829	0.97161	0.81590
	55	1.00000	1.00000	1.00000	1.00000	1.00000	1.00000	1.00000	0.99999	0.99912	0.98236	0.86437
	56	1.00000	1.00000	1.00000	1.00000	1.00000	1.00000	1.00000	0.99999	0.99956	0.98943	0.90333
	57	1.00000	1.00000	1.00000	1.00000	1.00000	1.00000	1.00000	1.00000	0.99979	0.99389	0.93339
	58	1.00000	1.00000	1.00000	1.00000	1.00000	1.00000	1.00000	1.00000	0.99990	0.99660	0.95569
	59	1.00000	1.00000	1.00000	1.00000	1.00000	1.00000	1.00000	1.00000	0.99996	0.99818	0.97156
	60	1.00000	1.00000	1.00000	1.00000	1.00000	1.00000	1.00000	1.00000	0.99998	0.99906	0.98240
	61	1.00000	1.00000	1.00000	1.00000	1.00000	1.00000	1.00000	1.00000	0.99999	0.99953	0.98951
	62	1.00000	1.00000	1.00000	1.00000	1.00000	1.00000	1.00000	1.00000	1.00000	0.99978	0.99398
	63	1.00000	1.00000	1.00000	1.00000	1.00000	1.00000	1.00000	1.00000	1.00000	0.99990	0.99668
	64	1.00000	1.00000	1.00000	1.00000	1.00000	1.00000	1.00000	1.00000	1.00000	0.99996	0.99824
	65	1.00000	1.00000	1.00000	1.00000	1.00000	1.00000	1.00000	1.00000	1.00000	0.99998	0.99911

Cumulative Binomial Distribution – 8

							p					
n	x	.01	.05	.10	.15	.20	.25	.30	.35	.40	.45	.50
100	66	1.00000	1.00000	1.00000	1.00000	1.00000	1.00000	1.00000	1.00000	1.00000	0.99999	0.99956
ctd	67	1.00000	1.00000	1.00000	1.00000	1.00000	1.00000	1.00000	1.00000	1.00000	1.00000	0.99980
	68	1.00000	1.00000	1.00000	1.00000	1.00000	1.00000	1.00000	1.00000	1.00000	1.00000	0.99991
	69	1.00000	1.00000	1.00000	1.00000	1.00000	1.00000	1.00000	1.00000	1.00000	1.00000	0.99996
	70	1.00000	1.00000	1.00000	1.00000	1.00000	1.00000	1.00000	1.00000	1.00000	1.00000	0.99998
	71	1.00000	1.00000	1.00000	1.00000	1.00000	1.00000	1.00000	1.00000	1.00000	1.00000	0.99999
	72	1.00000	1.00000	1.00000	1.00000	1.00000	1.00000	1.00000	1.00000	1.00000	1.00000	1.00000
	73	1.00000	1.00000	1.00000	1.00000	1.00000	1.00000	1.00000	1.00000	1.00000	1.00000	1.00000
	74	1.00000	1.00000	1.00000	1.00000	1.00000	1.00000	1.00000	1.00000	1.00000	1.00000	1.00000
	75	1.00000	1.00000	1.00000	1.00000	1.00000	1.00000	1.00000	1.00000	1.00000	1.00000	1.00000
	76	1.00000	1.00000	1.00000	1.00000	1.00000	1.00000	1.00000	1.00000	1.00000	1.00000	1.00000
	77	1.00000	1.00000	1.00000	1.00000	1.00000	1.00000	1.00000	1.00000	1.00000	1.00000	1.00000
	78	1.00000	1.00000	1.00000	1.00000	1.00000	1.00000	1.00000	1.00000	1.00000	1.00000	1.00000
	79	1.00000	1.00000	1.00000	1.00000	1.00000	1.00000	1.00000	1.00000	1.00000	1.00000	1.00000
	80	1.00000	1.00000	1.00000	1.00000	1.00000	1.00000	1.00000	1.00000	1.00000	1.00000	1.00000
	81	1.00000	1.00000	1.00000	1.00000	1.00000	1.00000	1.00000	1.00000	1.00000	1.00000	1.00000
	82	1.00000	1.00000	1.00000	1.00000	1.00000	1.00000	1.00000	1.00000	1.00000	1.00000	1.00000
	83	1.00000	1.00000	1.00000	1.00000	1.00000	1.00000	1.00000	1.00000	1.00000	1.00000	1.00000
	84	1.00000	1.00000	1.00000	1.00000	1.00000	1.00000	1.00000	1.00000	1.00000	1.00000	1.00000
	85	1.00000	1.00000	1.00000	1.00000	1.00000	1.00000	1.00000	1.00000	1.00000	1.00000	1.00000
	86	1.00000	1.00000	1.00000	1.00000	1.00000	1.00000	1.00000	1.00000	1.00000	1.00000	1.00000
	87	1.00000	1.00000	1.00000	1.00000	1.00000	1.00000	1.00000	1.00000	1.00000	1.00000	1.00000
	88	1.00000	1.00000	1.00000	1.00000	1.00000	1.00000	1.00000	1.00000	1.00000	1.00000	1.00000
	89	1.00000	1.00000	1.00000	1.00000	1.00000	1.00000	1.00000	1.00000	1.00000	1.00000	1.00000
	90	1.00000	1.00000	1.00000	1.00000	1.00000	1.00000	1.00000	1.00000	1.00000	1.00000	1.00000
	91	1.00000	1.00000	1.00000	1.00000	1.00000	1.00000	1.00000	1.00000	1.00000	1.00000	1.00000
	92	1.00000	1.00000	1.00000	1.00000	1.00000	1.00000	1.00000	1.00000	1.00000	1.00000	1.00000
	93	1.00000	1.00000	1.00000	1.00000	1.00000	1.00000	1.00000	1.00000	1.00000	1.00000	1.00000
	94	1.00000	1.00000	1.00000	1.00000	1.00000	1.00000	1.00000	1.00000	1.00000	1.00000	1.00000
	95	1.00000	1.00000	1.00000	1.00000	1.00000	1.00000	1.00000	1.00000	1.00000	1.00000	1.00000
	96	1.00000	1.00000	1.00000	1.00000	1.00000	1.00000	1.00000	1.00000	1.00000	1.00000	1.00000
	97	1.00000	1.00000	1.00000	1.00000	1.00000	1.00000	1.00000	1.00000	1.00000	1.00000	1.00000
	98	1.00000	1.00000	1.00000	1.00000	1.00000	1.00000	1.00000	1.00000	1.00000	1.00000	1.00000
	99	1.00000	1.00000	1.00000	1.00000	1.00000	1.00000	1.00000	1.00000	1.00000	1.00000	1.00000
	100	1.00000	1.00000	1.00000	1.00000	1.00000	1.00000	1.00000	1.00000	1.00000	1.00000	1.00000

Index

https://doi.org/10.1515/9781547401475-017

www.ingramcontent.com/pod-product-compliance
Lightning Source LLC
Chambersburg PA
CBHW060239220326
41598CB00027B/3982